In Memory Of

The Wiman Family

SOCIOPOLITICAL ASPECTS OF INTERNATIONAL MARKETING

Erdener Kaynak, PhD, Editor

SOME ADVANCE REVIEWS

This timely and unique collection fills a long felt need for a solid and thorough inquiry into the sociopolitical aspects of international marketing. The editor's judicious selection of the topics and contributors as well as the authors' masterful treatment of the studies make this book a definitive milestone. This should be of great interest to students and scholars of business, political science, international affairs, economics and sociology as well as to businessmen and investors who want to stay at the forefront of today's global business phenomenon.

M. Yasar Geyikdagi, PhD
Professor of Business & Management
State University of New York, Old Westbury

The environmental context is the heart of the matter in studying international marketing. In order to be helpful, a book on international marketing should not be based only on abstract concepts . . . it should be full of concrete examples in different countries, industries, and multi-national corporations. Based on this criterion [this] book is a definite success.

Kam-Hon Lee, PhD
Reader and Chairman, Department of Marketing
The Chinese University of Hong Kong

The book provides an original contribution to one dimension of international marketing that has seldom been treated, at that depth, before. The papers are well selected and well written. The geographical and sector coverage of the papers is especially interesting. The integration of the strategic dimension into the overall socio-political frame strengthens the analysis. In all, it is a solid contribution to the literature on the topic.

M. S. S. El-Namaki, PhD
Director, Netherlands International Institute of Management

NOTES FOR PROFESSIONAL LIBRARIANS AND LIBRARY USERS

This is an original book title published by The Haworth Press, Inc. Unless otherwise noted in specific chapters with attribution, materials in this book have not been previously published elsewhere in any format or language.

CONSERVATION AND PRESERVATION NOTES

All books published by The Haworth Press, Inc. are printed on certified ph neutral, acid free book grade paper. This conforms to recommendations ensuring a book's long life without deterioration.

Sociopolitical Aspects of International Marketing

Sociopolitical Aspects of International Marketing

Erdener Kaynak, PhD
Editor

The Haworth Press
New York • London

The Haworth Press, Inc. 10 Alice Street, Binghamton, NY 13904-1580
EUROSPAN/Haworth, 3 Henrietta Street, London WC2E 8LU England

Library of Congress Cataloging-in-Publication Data

Sociopolitical aspects of international marketing / Erdener Kaynak, editor.
p. cm.
Includes bibliographical references
ISBN 0-86656-951-0
1. Export marketing – Social aspects. 2. Export marketing – Political aspects. I. Kaynak, Erdener.
HF1416.S63 1990
658.8'48 – dc20 90-4532
CIP

CONTENTS

List of Tables

List of Figures

ABOUT THE EDITOR

Erdener Kaynak, PhD, is Professor of Marketing at the School of Business Administration, The Pennsylvania State University at Harrisburg. He has held key teaching and administrative positions in Europe, Asia, and North America.

Professor Kaynak has extensive consulting and advising experiences in international marketing in Europe, Latin America, the Middle East, and the Far East. He was Director of International Programs and served on the Board of Governors of the Academy of Marketing Science. In this capacity, he organized two World Marketing Congresses.

Dr. Kaynak has authored and co-authored many books and articles on international marketing and cross-cultural/national consumer behavior, some of which have been translated into Chinese, Spanish, Norwegian, Japanese, Hungarian, and Turkish. His latest book is *International Advertising Management.* Currently, Dr. Kaynak serves on several marketing journal review boards, and is Senior Editor for International Business, The Haworth Press, Inc. He is also editor of the *Journal of Global Marketing*, the *Journal of International Consumer Marketing*, the *Journal of Food & Agribusiness Marketing*, the *Journal of Teaching in International Business*, and the *Journal of Euromarketing*, all published by The Haworth Press, Inc.

Foreword

What kinds of knowledge and special expertise are needed to operate effectively in the incredibly complex global environment? Erdener Kaynak has given a great deal of thought to this question throughout his distinguished career as a scholar of international business. Professor Kaynak has always believed that to truly understand marketing in the global setting one must go well beyond superficial strategies and tactics of international marketing or the mere mechanics of establishing trading relationships and examine the *underlying processes* on which successful international marketing operations are based. It is in this spirit that his latest contribution to the international business/marketing literature is offered. *Sociopolitical Aspects of International Marketing*, a collection of thought provoking articles contributed by leading international business scholars from around the world, reflects Professor Kaynak's keen perception and skill at delving beneath the surface to select and organize an anthology that offers real insights into the environment of international marketing.

In Chapter 1, Professor Kaynak sets the tone for the entire volume with his penetrating yet succinct overview of the environment of international business. In a relatively few pages he provides the reader with a framework and integrative analysis of the international business environment unmatched anywhere in the international marketing literature.

Subsequent chapters focus on an incredibly diverse range of countries and their socio-political environments. What is especially noteworthy here is that these chapters do not focus on the often-examined environments of Western Europe and Asia but on more obscure yet critically important national environments all over the world. For example, the discussion of Poland in Chapter 8, Romania in Chapter 13, and Yugoslavia in Chapter 7 could not be more important and timely in light of recent developments in Eastern Europe. The same could be said for the material on South Africa discussed in Chapter 16 as well as the treatment of the socio-political environment of Grenada presented in Chapter 9. Other chapters such as the one on the Asia Pacific Region (Chapter 3), Egypt (Chapter 6), Nigeria (Chapter 14), Turkey (Chapter 18), and several others, while perhaps not as prominent in the recent news, are nevertheless of growing importance in the international arena.

While collectively the articles presented in *Sociopolitical Aspects of International Marketing* present a broad overview of some of the most vital yet neglected environments around the world, each individual article provides remarkable depth of coverage and analysis in the space provided.

Students of international marketing as well as serious practitioners will significantly enhance their understanding of the milieu of international marketing by reading this book. Professor Kaynak and all of the contributors should be commended for what I am sure will be viewed as an important contribution to the literature of international marketing.

Bert Rosenbloom
Drexel University

Preface

Whether we talk about domestic or international marketing, the marketing concepts and techniques that apply as marketing principles are universal. What is different in the international arena is the application of marketing principles. Despite great similarities between domestic and international marketing, there are also areas where there may be acute dissimilarities between the domestic and international markets. The area where these differences exist generally are pronounced in socio-economic, cultural and legal political environments. These are the areas where most of the international marketing blunders are observed. For effective marketing strategy preparation for international markets, decision makers of firms and public organizations need to study various aspects of the above mentioned environmental variables and make proper adaptation to them.

In view of this apparent trend, the purpose of this anthology of readings is to examine various facets of socio-economic, political and cultural aspects of international marketing management. To this end, topics of discussion are arranged in five sections. Section I is an integrative statement of the environmental issues in international marketing. Section II is related to macro international marketing issues where technology transfer issues, overcoming trade barriers in the Asia Pacific region and government support for export entry strategy are discussed. Section III examines very critically various socio-political issues in international marketing by use of specific foreign markets as case examples. In particular, socio-political marketing issues and strategies related to company, government and consumer groups involving countries as diverse as Holland, Egypt, West Indies, Poland and Grenada are examined. Section IV is a treatment of various marketing decision variables in the adaptation to changes taking place in international marketing environment. Finally, in Section V special international marketing topics such as social responsibilities of marketing, husband and wife decision making in an oriental culture and tourism marketing issues are examined both conceptually as well as empirically.

During the preparation of this volume, the editor has received the help of many individuals too difficult to name one by one. I would like to thank the contributors for their conscientious efforts to deliver their individual chapters on time. At the Pennsylvania State University at Harrisburg, I received the help and encouragement of Dr. Melvin Blumberg, Director of the School of Business Administration for which I am grateful. Word processing help came from Linda Zubler and Sharon L. Kreider, whose help was critical. Last but not least, my family was again a constant source of help and encouragement. Of course, without their sacrifices it would have been far more difficult to complete the project on time.

Needless to say that I am solely responsible for any errors or omissions in the volume.

Erdener Kaynak
Hummelstown, Pennsylvania

SECTION I: INTRODUCTION

Chapter 1

Environmental Issues in International Marketing: Integrative Statement

Erdener Kaynak

INTRODUCTION

Social, economic, political, and cultural environmental factors affect the international marketing performance and operation of firms in multiple environments. The environment contains certain conditions that limit the scope of activity of the international marketing company and affect its organizational structure. The overall pattern is such that environmental factors shape and set limits on entry and operating decisions of international companies of different sizes and coming from varying countries and industries/sectors. Environmental factors, such as income level, cultural attributes, consumer characteristics, and legal and political environments, have tremendous bearing on the market potential of international firms and thus the scale of those companies' operations.

Among the international environmental factors, the most significant ones are the social and political, compared to a domestic marketing environment. These two major environmental factors force international companies to adopt corresponding marketing techniques and practices in different international markets. Marketing institutions, methods, and techniques evolve with a changing social structure which, in turn, has evolving political, legal, economic and business components. In international markets, differing combinations of social, economic, legal and political conditions create differing marketing systems. Thus, the marketing system changes with changes in its international environment.

The comparative approach to marketing systems involves the systematic detection, identification, classification, measurement and interpretation of similarities and differences of various distributive institutions operating in foreign markets (Boddewyn, 1965). It is also necessary to discern what is universal, related and unique among the varying marketing systems of diverse nations (Kaynak, 1980a).

It must be stated here that not every facet of a country's marketing system is significantly affected by the numerous elements in its socio-economic, demographic, legal-political, technological and cultural environment. For instance, in the case of channels of distribution, it would be wise to use a few, related variables to examine the various distribution systems and institutions (Shapiro, 1965), since the focus of such a study is on the systems and not on their environments.

An environment refers to what is external to a marketing system, and neither directly controls it nor is directly controlled by it. The effects of the various elements of the environment on marketing and its components are explored conceptually and analytically in other studies (Carmen and March, 1979; Hakanson and Wootz, 1975). As a rule, comparisons of marketing systems are made across national and cultural divisions. Inter-cultural and inter-regional comparisons seem most natural for comparative marketing and distribution analyses (Green and Langeard, 1975; Harris, Still and Crask, 1978; Hoover, Green and Saegert, 1978).

The areas of marketing in which comparisons have been made at the international level have been concerned with the channels of distribution, especially those of wholesaling and retailing. The study of distribution treats the differences in distributive practices in various foreign markets as being as important as the similarities. The critical element in comparative distribution in overseas markets is the manner in which experience gained in the U.S.A. is interpreted, related and generalized.

Most marketing scholars and practitioners have interpreted the prevalent marketing practices in terms of socio-economic conditions of a foreign country (Bartels, 1963). This author does not want to undermine the effect of socio-economic factors, but considers other environmental factors. For example, the supplier environment, competition and governmental legislations and actions may also affect the characteristics of foreign country marketing systems and strategies. The causal relationships between a developing country distributive system and its various environments are illustrated in Figure 1.1.

The theory that the marketing system of a country is closely related to the development of its social, economic, technological and cultural environments is widely held by many scholars (Wadinambiaratchi, 1965; Kaynak, 1979). However, the precise nature of the relationship between environmental factors and the marketing systems is a matter of speculation (Douglas, 1971). A marketing system of a country seeks to satisfy human needs, but the manner in which it performs its functions varies among different societies. It can be hypothesized that a pattern of marketing practice, in terms of the types of institutions, marketing practices, organization of firms, managerial attitudes, strategies utilized and channel structure, may be expected to emerge within a variety of countries under the influence of various environmental conditions (see Table 1.1).

MARKETING SYSTEMS IN FOREIGN MARKETS

A marketing system in a foreign country environment is an aggregate abstract concept; the basic units or organizations of such a system are the marketing institutions whose individual actions result in changes in the marketing system. They are also influenced by the workings of the system. Similarities among marketing systems of foreign countries exist not only in the case of modern institutions, such as supermarkets, but also in traditional institutions. These similarities are due to the same forces and processes operating on the environment (Goldman, 1970).

The organization and structure of the market in these countries is a function of certain economic variables, such as the size of the market, the nature of the goods, the size of the production units and the degree of specialization. These variables show a basically similar pattern of development throughout the densely populated countries of the world because they are the result of a certain stage of

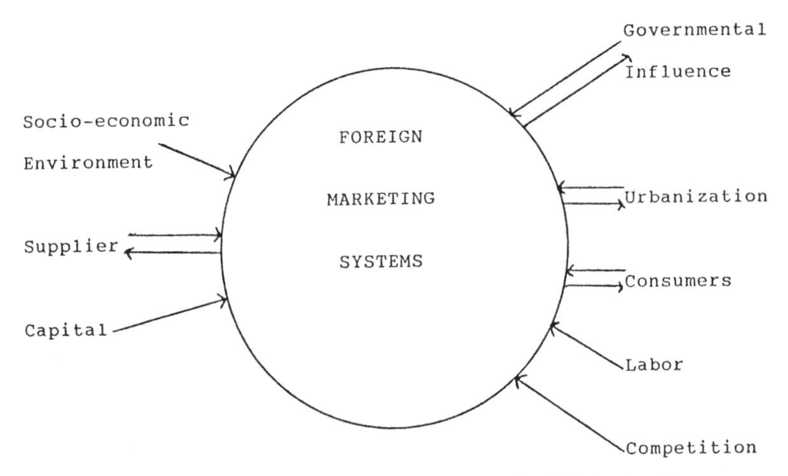

FIGURE 1.1 Environmental Pressure on Foreign Marketing Systems

economic development (Cunningham, Moore and Cunningham, 1974; Kaynak, 1986). The interaction taking place between the marketing system and the surrounding environment is shown in Figure 1.2.

In order to have a better grasp of the marketing system in a developing country, it is necessary to discuss factors which tend to create marketing problems or influence distribution methods, namely: the geography of the country, market factors, and concentration of industries (Darling, 1966). The continuous growth of a market has provided an opportunity for new forms of marketing institutions to take over an increasing share of the market, without displacing traditional operations. Today, developing countries of the world are characterized by a continuous increase in income and population density. This constitutes an ever-present source of pressure for the development of new forms of distributive institutions.

In most cases, the complexity of the marketing system is related to the stage of a country's socio-economic and technological development. In a developed economy, there are highly integrated manufacturing, wholesaling and retailing functions. On the other hand, in developing countries, there is a lack of integration among various functions at each level (Goldstucker, 1968, p. 470). The conditions under which the marketing process functions in a developing econ-

TABLE 1.1. Comparison of Marketing Systems in Foreign Markets

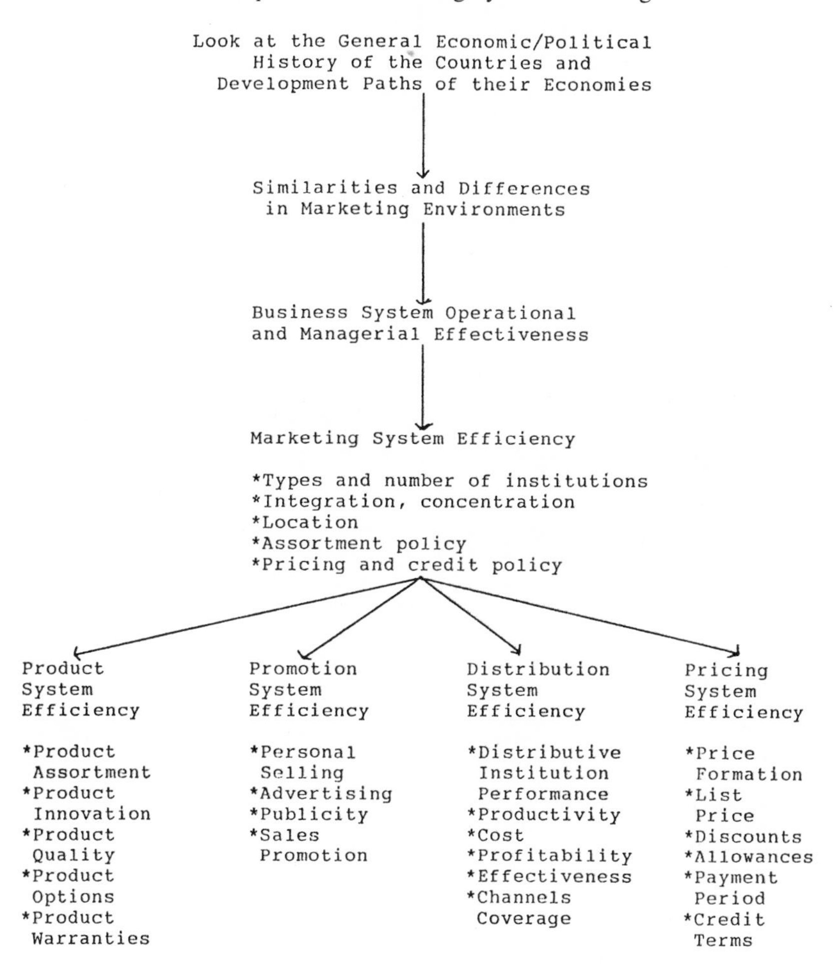

omy such as Jamaica have created many different types of distributors: some specialist middlemen, some producer-distributors. The characteristics of the items handled in terms of bulk, unit price, perishability, fragility, seasonality; the character and amplitude of the demand; the seasonal and regional variation in supply; and the means and costs of transport all influence the form which the distribution of a given product will take (Mintz, 1956).

FIGURE 1.2. Interaction Between Marketing Systems and the Environment

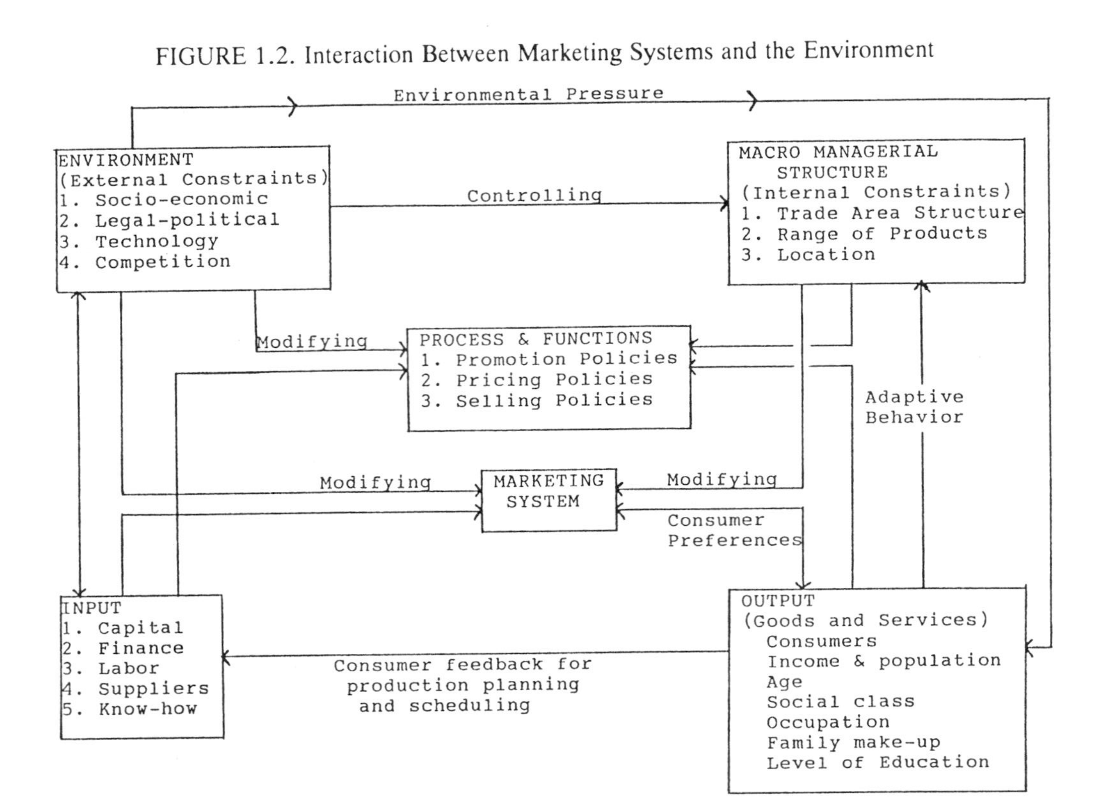

Source: Erdener Kaynak, "A Refined Approach to the Wheel of Retailing," _European Journal of Marketing_, Vol. 13, No. 7, 1979, p. 242.

In developing countries, distributive institutions and establishments are typically small in size. This means that the turnover of these outlets is low, their facilities are restricted and the goods they handle are limited. Whereas large scale distributive institutions are mainly concentrated in major cities (Guerin, 1965), entry into the distributive system is relatively easy because of the low investment and technical skills required. The result is multiplicity of small scale operations characterized by poor management practices. These stores tend to operate with a minimum of space and the scale of operations in most cases is low. For this reason, the distributor seeks a high margin rather than a high volume to survive (Izraeli, 1974). Despite its many disadvantages, the present system of distribution in developing countries maintains its existence because it provides several advantages to its different constituencies. It may act as a form of unemployment relief in public sector enterprises.

In developing countries, certain individual traders have very little capital, and they cannot afford to deal with such a long and expensive distribution process independently. As a result, each trader forms a link in a long chain of intermediaries, making only a small profit, but using relatively little capital in their transactions (Bromley, 1971). Forman and Riegelhaupt (1970) suggested that, in the early stage of commercial development, distribution chains become gradually longer and small-scale intermediaries proliferate. At a later stage, increasing demand and bulk transport facilities lead to the shortening of the distributive chain as the small-scale, mainly rural intermediaries are undercut and by-passed by large, urban wholesalers, using a high input of capital and having turnover but a relatively low profit margin. Intricate networks of channel intermediaries facilitate distribution over long distances, and the existence of storage facilities and a long tradition of internal trading ensures harmony between supply and demand of a variety of products. Generally, spatial and temporal arbitrage take place, but traders and consumers nevertheless are quite responsive to price variations. Most of the intermediaries are quite specialized, and there are several different kinds of actors involved in the domestic marketing process (McKay and Smith, 1973).

It is pointed out that the small scale food producers in developing countries require much more technical marketing assistance before they can meet the requirements of a modern integrated food marketing system (McKay and Smith, 1973, p. 8). At the intermediary level, namely wholesaling and retailing, careful economic analysis is needed. Low income of the consumer limits the assortment of the retail store and the customer's outreach. Small stores located near the consumer homes seem, therefore, essential for convenient daily food purchasing. Since labor costs are low, the application of labor-saving equipment should have low priority. Taking these basic differences, namely low labor costs and low consumer incomes into consideration, the scope for marketing system development in developing countries should be on the following focal points (FAO, 1975).

a. How to improve and strengthen the services of marketing system participants for the benefit of consumers, producers and other participants in the marketing chain to reach the major objective of a marketing system to become a major dynamic force for economic development.
b. Reduction of costs through improving marketing system.
c. Form of marketing structure, namely what is the most suitable type of marketing institution to be promoted and what should be the most suitable distribution structure for it.
d. In view of the important role some products play in the family budget, particular attention should be given to improving their marketing system, particularly under conditions of shortage.
e. Given the present shopping behavior of developing country consumers, how to develop an efficient marketing system which can not only satisfy the needs of low income consumers, but also offer reasonable return without being a burden on taxpayers.

The total production, marketing, processing and distribution functions for different food products in a developing country can be viewed as a "commodity system" of linked steps through which the products flow to final customers. It is of paramount importance to understand the interfaces between production and marketing for each group of products as a commodity system. This necessitates systematic planning and development of all vital activities which are necessary to effectively produce, process, market, distribute and utilize a commodity. Generally, food system effectiveness in developing countries is measured in terms of how all components of the food chain function together. To this end, the commodity system approach can serve as an important tool in planning, implementing and evaluating commodity-oriented food system development activities.

Trade, especially in the Asia Pacific region, is of vital importance to the health of world economies. Various barriers exist which impede the free flow of products across national boundaries. The more common barriers are those in the form of tariffs, and quantitative restraints in the form of quotas. However, other barriers exist in the form of domestic protection measures; and even because of esoteric differences in culture or political philosophies.

It is stated by Patricia Dunn Duffy that global businesses must encourage freer trade rather than promoting domestic protectionist measures in order to ensure long-term survival. A common reaction by a nation whose products are denied importation by another country is the retaliatory imposition of similar restrictions. What ultimately results is that both countries suffer dramatic economic consequences. This can lead to political tensions, and even war.

While many barriers to trade abound, businesses and individuals can profoundly reduce their impact. Multilateral cooperative trade agreements between nations, or through business negotiations, can greatly reduce or even eliminate barriers to trade. Additionally, many impediments to product exchange can be diminished through exploding popular misconceptions about economic welfare and trade.

The better business and marketing professionals understand the social and political implications of trade barriers, the more effectively they can formulate appropriate international marketing strategies and overcome impediments to global product exchange. The greater the cooperation between firms and nations, the healthier the world economy and the chances for eliminating international tensions.

The purpose of the study by Hyder and Ghauri is to describe the process of technology transfer through a joint venture relationship with a longitudinal perspective. The dynamic character of the relationship has been more or less ignored previously, so it has become important to comprehend how it develops over time. To collect the necessary information, both partners were interviewed several times to follow the development of the relationship. A model is developed so that the study can be systematically described and analyzed. Interaction between exchange of resources, control and co-operation/conflict constituted the relationship, which is influenced by partners' interests. Performance of the activity will have a decisive affect on how much interest partners should attach with the joint venture. A major part of the study has been devoted to the comparison of the expected contribution at the initial stage with the actual contribution during the operating period. An attempt is also made to outline general as well as specific implications from the experience of the case study.

The chapter by Rolf Seringhaus addresses the export market entry situation from the perspective of the experienced exporter by focusing on decisions, entry problems and the role played by governmental support. A great deal of attention has been devoted to the process of increasing export involvement, or export stages, and the recognition that the early phases of such a process appear to be most critical to future export behavior of the firm. The study reported in this chapter examines the decision parameters involved, and the actual entry difficulties encountered by firms in entry of a new export market, and where government support can contribute to this process.

A group of ninety small-and medium-sized Canadian exporters provided data in a structured personal interview on their most recent overseas market entry. Decision facets that played the greatest role in this market entry were the firms' profit expectations and management interest in the venture. Overall, however, it was a group of variables best described as "effort that reflects the importance of decisions." The most problematic aspects of entry concerned competition and the access to the new market. There seemed to be another dimension to entry problems, namely that of "expertise." Government support is extensively used for market contact and market access. It is in these areas that support programs contribute most to export marketing. While firms' knowledge of markets benefits in the process, their preparation for export and their actual marketing competence in a new market, however, seems only marginally affected by governmental support. The relationship between decisions and problems showed that firms emphasized commitment and attitude on the one hand, and encountered problems in the market environment and dynamics on the other.

It is apparent that exporters must accept a continual learning mode, that is, new markets pose new challenges that cannot be fully met with old knowledge and expertise. The key implication is that exporters' marketing competence is indeed dynamic and that government support used selectively and in a focused way, can yield benefits to the firm. Management action to cope is threefold: first, review past entries and learn; second, plan and prepare realistically for new entry; and third, assess the firm's competence pragmatically and objectively.

Managers are faced with unfamiliar change, with finding new answers to new questions. Do corporations have to respond or not, why, when, how much and how? How do managers deal with this change organizationally, inside and out?

The chapter by Nico J. Vink deals with the way in which managers are learning the new and unlearning the old. It shows us how they are learning from others, how they set examples, how to discover new types of action and new ways of interacting—just do it and become company heroes. How they motivate others to discover the new, create "safe fail" corporate cultures in which others are willing to experiment and change the rules of the game in action-oriented ways.

Management should ask itself a number of important questions. Have I done everything I should in learning myself and in teaching others to: (a) stop the situation of "orphan strategy" and "passing batons" and to become really interactive instead; (b) understand the way we do things here, the constancies as well as the fluctuations, and why that is so; (c) become rightly responsive, not to complicate or oversimplify and not to overreact by doing too much or too little; (d) be socially and societally intelligent; and (e) to integrate strategy and implementation, and to tap and mobilize the company's responsiveness to culture and power culture to do the right things right?

The socio-political environment is the non-economic environment which encompasses those elements of human society that manifest themselves in political change. Political change is important for the international marketer when it results in the implementation of policies that affect trade and investment. The focus of the chapter by Mahmoud and Rice is the socio-political environment in Egypt. First, a general and brief examination of the nature of the socio-political environment as it pertains to international marketing is given. Three socio-political analytical frameworks are then described: Gillespie's coalition cycle, bargaining power analysis, and de la Torre and Neckar's approach to political risk forecasting. The frameworks are illustrated and evaluated using Egypt's experience with "infitah," its economic liberalization policy. The evaluation highlights some weaknesses in the frameworks. One example is the variable measurement problem. Yet, the frameworks make a useful contribution to an international marketing manager's attempts to understand the socio-political environment and formulate strategies accordingly. Successful socio-political analysis takes a dynamic approach—it is proactive rather than reactive—and is conducted at macro (country) and micro (firm or project) levels.

The chapter by Renwick and Renwick examined the social linkage effect upon perceived product attributes from exporting nations as reported by importers with purchasing experiences with these same countries. The importers sampled constituted a diverse sectoral representation from five Caribbean countries: Trinidad, Jamaica, Barbados, Bahamas, and Haiti. Their perceptions, purchasing experience, and social linkages (friendships) were examined with respect to the exporting nations that were the major trading partners in the region (i.e., United States, Canada, United Kingdom, France, Italy, West Germany, Netherlands, and Japan).

Previously reported research has demonstrated the efficacy of the social linkage effect upon country of origin stereotypes. A consistent positive effect of social linkages has been demonstrated upon the country of origin stereotype regardless of the purchasing experience with the exporting nations. The research reported here tested the effect of social linkages upon a set of salient product attributes (i.e., product quality, service, availability, delivery, and price) which collectively represent three levels of abstraction on the continuum of knowledge concerning product attributes. These attributes were identified by means of an elicitation survey conducted with American and Canadian importers. The results of this research indicate that the social linkage effect is strongest for product quality, that is, at the most abstract level on the product attribute continuum.

The implications of these findings for international marketing strategy are discussed. The discussion includes: attention to advertising, marketing opportunity analysis, export entry marketing, competitive strategy, and product level decisions for international markets.

The last few years have brought an unusual dynamism to the international exchange of technology process. From China to Romania, almost every communist country has implemented new regulations related to administration of technology transfer. Joint ventures with foreign capital is a new strategic resource for most communist countries. It is the way to generate new capital to pay their huge debts. The research issue discussed by Lis and Sterniczuk is how communist countries are to participate in joint ventures with foreign capital without being exposed to capitalism in their own country. In particular, the political and economic conditions of joint ventures with foreign capital in a communist system is examined.

The chapter by Lionel Mitchell examines politics and marketing in Grenada and the implications of a coercion factor. Grenada's welfare, growth and survival as a small developing country and a free enterprise system democracy are affected to a significant degree by political and other external factors and marketing. Grenada operates today in a changed and changing, complex, and more difficult environment compared to a decade ago. The blending and degree of importance of each of the elements of the total environment could exacerbate Grenada's problems and difficulties. This would be because of its poverty, small size, low political importance, low development, low per capita gross domestic product (GDP), and the demonstration effect in the context of its social, economic and political goals and aspirations. Misunderstanding or insensitivity (by governments and their agents) to its culture and other factors in the planning, development and implementation of marketing and economic programs could do unnecessary and great harm to the people and the country.

The distribution system for fresh, frozen and processed seafood products in the Atlantic Canadian province of Nova Scotia is analyzed by Kaynak and Rice by using the political economy framework. They describe the organizations, facilities and movement of seafood products from the production point to the final consumer. The first section discusses the internal political economy of the distribution channel; the second section gives an overview of the external political economy of the channel environment.

Theoretically, international marketing is based on "rational" decisions, usually in the sense of seeking acceptable goals and evaluating alternative ways of achieving them before taking any course of action. Bedford A. Fubara examines socializing grain distribution in a state of Nigeria. Grain State Enterprise in Kongi State has decided in a reactive manner to sell its grain internationally because of pressures to pay for its imports. The socio-political environment is such that the state economy is under the control of the government to generate surplus from economic activities for the improvement of the quality of life. The country is undergoing social transformation, having the backing of the working people, in a revolution against pseudo-capitalism.

The economy of the state is poor with a per capita GDP of US $110. It is an agrarian economy with poorly developed infrastructure. More than 80% of the population depends on agriculture for their livelihood. The Grain State Enterprise is set up to socialize grain distribution and to "wither-away" competition. It is expected to sell grain to the public enterprise engaged in export trade (at 50% of its purchase price).

GSE operation is expected to "eliminate" competition because it has to buy grain at collection points in all the districts of Kongi State. GSE purchase strategy is supported by the national task force to ensure the supply of grain from co-operative societies at the government's fixed price. Private suppliers are allowed to sell 50% of their production in the open market while state farms are sanctioned to sell 100% to GSE.

The philosophy of socialization is to expand the share of the socialist sector in the wholesale trade and to avoid duplication of efforts and overlap of functions in the economy. The implication of the socialization strategy in the international market is that GSE may assist the government to generate foreign exchange, but the strategy seems to create poor productivity, poor quality of grains, and poor motivation for grain farmers because of the negative pricing system. The strategy would be shortlived because it is self-defeating.

Pricing decisions have always been of great managerial concern, particularly in global and mature markets. The chapter by Martenson shows that IKEA, probably the world's largest furniture retailer, has choosen to keep prices low all over the world at the same time as it adapts its price level to the local markets. As a global company, IKEA has the opportunity to have lower prices on markets of strategic importance if competitive pressure makes it necessary, and to have higher prices on smaller, less important markets where the competitive pressure is much lower. This strategy gives the largest worldwide profit, which is the planning horizon of a global firm. It should be emphasized, however, that it takes some form of competitive advantage to be able to use price as a strategic tool. In addition, most companies are not global and have far less freedom to choose the pricing policy they would like.

The review of earlier studies on pricing also shows that it is much more difficult to standardize prices than products, for example, across countries. The empirical analysis of IKEA, a global retailer's pricing in eight countries, points at the advantages with operations on several markets. This study also points at the lack of theoretical development in the international pricing area. Is a standardized pricing approach of a low price image on all markets more appropriate, or does it mean the same prices across markets? IKEA has chosen a low price image on all markets, but prices on individual items sold do vary from one market to another.

The Romanian experience in central planning of channels of distribution is presented by Jacob Naor. The process, while highly bureaucratized, appears to permit rapid structural changes and is thus of relevance to countries in need of such changes, as well as to countries facing persistent resource constraints. In the Romanian context, distribution aims at achieving equity goals as well as cost efficiency, while meeting "rational" consumer needs. The planning process proceeds this from the determination of "rational" consumption requirements and the determination of retail and wholesale capacity needs, to output needs.

While rapid structural changes as well as distributional equity and some distribution-related cost savings appear to have largely been achieved, the system may nevertheless be presumed to suffer from overbureaucratization and diseconomies of scale, affecting retail service levels primarily. Incentive and initiative related to distribution appear to be severely limited and planners' prerogatives no doubt supercede consumer prerogatives. The attractiveness of the Romanian model for other change-oriented economies will thus depend largely on the ability of Romanian planners to mitigate such dysfunctional phenomena, while maintaining their ability to effect structural changes in an equitable manner.

Traditionally, interest by western students in marketing developments in Socialist countries has been sporadic at best. With some notable exceptions (Goldman, 1963; Naor, 1982; Noar, 1984) few serious attempts have been made to assess the impact of centralized planning on the character and structure of evolving marketing systems in socialist systems. This apparent lack of interest is surprising in light of the strongly established position of planning in Western marketing thought and practice. Indeed, one is hard pressed to think of the two independently, given the strongly managerial orientation of much of marketing thought in the West.

While it is true that marketing planning in the U.S. has been largely confined to the micro level, the experience of centrally planned macro systems is nevertheless relevant. Users of such systems have, for example, been able to bring about rapid structural changes, thus providing for rapid economic development. Romania is a prime example in this regard. However, to what extent have channels of distribution been part of such structural changes, and what, in general, appears to be the conceptual structure underlying distribution in highly centrally planned systems?

The chapter addresses such issues by presenting the prototype of a highly centralized and bureaucratized distribution system prevailing in the East-bloc, that of Romania. In an era of increasing resource constraints and growing government intervention, the operations of centrally administered systems such as that of Romania, appear to be an area that deserves careful attention.

The chapter by Julius O. Onah examines one of the contentious issues in marketing, namely; ethics and social responsibilities of marketing with particular reference to Nigeria. With the increase in both the standard of living and the level of education, consumers are now asking questions about the behavior of the businessmen in the marketplace. There is now, more than ever before, the demand for marketing ethics and social responsibilities of marketing.

In that chapter, ethics is defined as the philosophy of moral norms and values. It is concerned with human conduct of right or wrong. Examples of unethical behaviors in Nigerian business now abound. Such behaviors as passing blame for errors to an innocent co-worker, claiming credit for someone else's work, and pilfering company materials and supplies were among those considered to be enemical to business practice in Nigeria.

In tracing the evolution of marketing ethics, four major stages of sensitivity are highlighted. These are: (a) Ethics of self-interest; (b) Ethics of compulsion; (c) Ethics of compliance; and (d) Ethics of conviction.

There have been a lot of arguments about whether or not there is ethics in marketing. This chapter identifies major sources from which marketing ethics would be drawn. It also proposes a model for social responsibilities of marketing. Toward the end of the chapter, business actions toward socially responsible marketing are discussed to alert the manager who wants to achieve socially responsible marketing in his company. The Nigerian Marketing Association (NIMARK) Guides to professional conduct are provided as Appendix 14.1 to the chapter.

The chapter by Yau and Sin attempts to investigate husband-wife influence at different stages of the family purchasing decision process and to see if differences in perception exist between working wives and non-working wives in family purchasing decisions in an oriental environment. In order to achieve the objective, a sample of 75 couples was drawn by the multi-stage area sampling technique. The personal interview method was adopted to obtain systematic information from spouses in each sampling unit. Respondents were requested to indicate the major influence on twenty household purchasing decisions/products in different decision making stages. Then the multidimensional scaling technique was employed to obtain perceptual mappings of family purchase decisions from both husbands and wives. Findings showed that husband-wife influence at different stages existed, and their perceptions of various purchase decisions were consistent across different stages. Results also supported a shift from wife-dominant purchasing decision making to joint decision making in the case of working wives.

It is stated by Marius Leibold that the strategic stakeholder approach to the marketing of tourism and recreation facilities involves a new orientation to the unique complexities of marketing tourism/recreation facilities and destinations. The challenge for a particular tourism destination is to achieve an optimum fit between the satisfaction of tourism and recreation needs, utilization of tourism potential, and balanced local infrastructural development and community benefits.

An empirical research project investigated the existing approaches to marketing of tourism and recreational facilities by local public authorities. Four approaches were determined, viz. product, selling, consumer needs and community interest approaches. These approaches provide partial solutions, with none integrating the interests of all stakeholders in a local tourism industry.

The nature and task of contemporary marketing, in a tourism context, are examined. The uniqueness of tourism marketing, in terms of tourism "products," tourism "intermediaries," and the nature of tourism demand and supply are highlighted to serve as the foundation for development of the new approach.

The strategic stakeholder approach is based on two dimensions: (a) environmental behavior, both external and internal, and (b) organizational processes, both macro and micro. Underlying this basis is the socio-economic paradigm in marketing, which is a major departure from the micro-economic paradigm underlying the existing approaches. The strategic stakeholder approach utilizes constructs from exchange theory, organization theory, political science and sociology. The objective of its application is to achieve an optimum reconciliation of the aims, interests and actions of all tourism/recreation stakeholders of a local community.

During the main tourism season (June 15-September 15) in 1987, the Institute for Tourism, Zagreb and the Committee for Tourism of SR Croatia carried out a survey on the Attitudes and Expenditures of Foreign Tourists—TOMAS '87. The main goal of this research effort was to gain insight into the principal characteristics of the foreign tourism demand in SR Croatia. Application of such data about the market makes it possible to evaluate the tourism supply elements and it provides possibilities for tourism market segmentation according to various characteristics. The work presented by Weber and Telisman-Kasuta examines the results of a market segmentation study based on the respondents' attitudes toward 23 elements of Croatian tourism supply combined with their socio-demographic characteristics.

The factor analysis has, based on consumer attitudes, grouped particular supply elements (entertainment, sports, shopping and other facilities, basic supply elements, environmental elements, suitability for family vacation), at the same time showing both the respondents' perception of the destination features and pointing to the specific types of tourism supply.

The discriminant analysis was applied to differentiated respondent groups according to their attitudes toward the tourism supply, showing the age groups and country of permanent residence as the most discriminative features. Further, using the results of both multivariate methods, possibilities are suggested for evaluating the tourism supply in relation to the target groups. Such a manner of segmenting the foreign tourism market will enable the Croatian tourism industry to communicate with the chosen market segment(s), to develop a tourism product attuned to their wishes, needs and interests and in that way to be more competitive in the international tourist market.

Tourism, a labor-intensive service industry, tends to generate employment opportunities across a broad spectrum of job skills and levels of education. It may also have spill-over effects in other sectors of the economy such as agriculture, food processing, and light manufacturing, as well as engender support for traditional handicraft industries. Furthermore, a strong tourist industry can provide an important source of foreign exchange.

In developed countries, tourism is profitable, and earnings from tourism have been steadily increasing. However, the developing nations neglect tourism marketing, and an important source of income in most cases is ignored. As a result, tourism revenues in developing countries fall well below the tourism income of developed nations.

A major problem in establishing a tourism industry in these countries is the large initial investment required. In developing nations, the practical approach is to enhance opportunities for joint ventures with multinational corporations specializing in tourism. These companies provide not only the initial capital, but also marketing expertise training for management and support staff. If joint ventures can help find a tourism industry, then developing nations must first attract joint ventures, accommodating the needs of major development projects (for example, land acquisition) and gauging types and level of tourism demand.

Generally there are three groups of travelers: (1) high income, (2) middle income, and (3) low income: and four distinct market segments; (1) allocentrics, (2) near-allocentrics, (3) mid-centrics and (4) near-psychocentrics. Members of each group and segment have different demographic and attitudinal characteristics which inevitably determine the total tourism potential in a particular country. The chapter by Yucelt and Isley examines the tourism possibilities in the Mediterranean basin as a whole, focusing on a forecasting methodology which can be useful for estimating the level of expected demand. It uses Turkey as a case example and suggests strategic alternatives for Turkish tourism planning in light of the competition from Italy, Spain, France, Yugoslavia, Greece, Tunisia, Morocco, and Egypt. Over the years, tourist demand for Turkey has fluctuated erratically; for example, during the period of political turmoil and social unrest from the mid- to late 1970s, all economic activity suffered.

The empirical analysis consists of two different tests. The first is a Box-Jenkins time series test to determining if there has been a structural change in the generating of tourists' demand for travel to Turkey. The second is a market share analysis to determine to what extent, if any, Turkey's market share of tourists' demand for travel to the Mediterranean basin has changed.

The data for the Box-Jenkins analysis are number of arrivals of tourists by month from 1958 to 1985. This model was fitted to the entire series and to a subperiod within the series. The findings yield no evidence that the disruption of the 1970s of that subsequent policy has had any effect on the development of demand for travel to Turkey.

The data for the market share analysis covered a span of twenty-one years, 1961-1981, for twelve countries in addition to Turkey: Spain, France, Italy, Greece, Israel, Yugoslavia, Egypt, Algeria, Morocco, Cyprus, Malta, and Tunisia. An analysis of the correlation (covariance) showed strong positive or direct relation between the Turkish market and market shares for Spain, Greece, Egypt, Morocco, and Tunisia. Travel to these countries might be viewed as complements to travel to Turkey. However, an inverse relationship existed between the Turkish market share and the market share for France, Italy, Israel, Algeria, Cyprus, and Malta. These countries attract tourists for whom travel to Turkey seems to be a substitute.

To be competitive in the Mediterranean basin, Turkey should: (1) recognize that there are a large number of substitutes for travel to Turkey; (2) offer alternative tourism facilities such as sport facilities and dining, sightseeing, and shopping opportunities; (3) educate local residents on how to assist visitors; and (4) develop a comprehensive tourism marketing strategy and associated advertising plan for immediate use. This strategic plan should be integrated into the country's Five Year Economic Development Plan for better resource allocation to this neglected but very important industry.

REFERENCES

Bartels, R. (1963). *Comparative Marketing: Wholesaling in Fifteen Countries*. Richard D. Irwin, Inc., pp. 1-6.

Boddewyn, J. (1965). "The Comparative Approach to the Study of Business Administration," *Academy of Management Journal*, Vol. 8, December, pp. 261-267.

Bromley, R.J. (1974). "Interregional Marketing and Alternative Reform Strategies in Ecuador," *European Journal of Marketing*, Vol. 8, No. 3, pp. 245-264.

Carment, J.M. & March, R.M. (1979). "How Important for Marketing are Cultural Differences Between Similar Nations?" *Australian Marketing Researcher*, Vol. 3, No. 1, Summer, pp. 5-20.

Cunningham, W.H., Moore, R.M. & Cunningham, I.C.M. (1974). "Urban Markets in Industrializing Countries: The Sao Paulo Experience," *Journal of Marketing*, Vol. 38, April, pp. 2-12.

Douglas, S.P., (1971). "Patterns and Parallels of Marketing Structures in Several Countries," *MSU Business Topics*, Vol. 19, No. 2, Spring, pp. 38-48.

Forman, S. & Rieglehaupt, J.F. (1979). "Market Place and Market System: Towards a Theory of Peasant Economic Integration," *Comparative Studies in Society and History*, Vol. 12, p. 202.

Goldman, A.(1970). "Developments and Change in Retail Systems," Unpublished Ph.D thesis, Berkeley: University of California.

Goldstucker, J.L. (1968). "The Influence of Culture on Channels of Distribution," in King, R.L. (ed.) *Marketing and The New Science of Planning*, AMA 1968 Fall Conference Proceedings Series No. 28, pp. 468-473.

Green, R. & Langeard, E. (1975). "A Cross-National Comparison of Consumer Habits and Innovation Characteristics," *Journal of Marketing*, July, pp. 34-41.

Guerin, J.R. (1965). "The Introduction of a New Food Marketing Institution in an Undeveloped Economy: Supermarkets in Spain," *Food Research Institute Studies*, Vol. 5, No.3, pp. 217-227.

Hakanson, H. & Wootz, B. (1975). "Supplier Selection in an International Environment," *Journal of Marketing Research*, February, pp. 46-51.

Harris, G., Still, R. & Crask, M. (1978). "A Comparison of Australian and U.S. Marketing Strategies," *Columbia Journal of World Business*, Summer, pp. 87-94.

Hoover, R., Green, R. & Saegert, J. (1978). "A Cross-National Study Perceived Risk," *Journal of Marketing*, July, pp. 102-108.

Izraeli, D. (1974). "Priorities for Research and Development in Marketing Systems for Developing Countries," Working Paper No. 209/74, April, Tel Aviv University.

Kaynak, E. (1986). *Marketing and Economic Development*, New York: Praeger Publishers Inc.

Kaynak, E. (1980). "Future Directions for Research in Comparative Marketing," *The Canadian Marketer*, Vol. 11, No. 1, pp. 23-28.

Kaynak, E. (1979). "A Refined Approach to the Wheel of Retailing," *European Journal of Marketing*, Vol. 13, No. 7, pp. 237-245.

McKay, J. & Smith, R.H.T. (1973). "The Role of Internal Trade and Marketing in Development," In *Development of Tropical Environments Symposium*, 45th Congress, Perth, Australia, August 13-17, p. 15.

Mintz, S.W. (1956). "The Role of the Middleman in the Internal Distribution System Caribbean Peasant Economy," *Human Organization*, Vol. 15, No. 2, Summer, pp. 18-23.

Shapiro, S.J. (1965). "Comparative Marketing and Economic Development," In Schwartz, George (ed.) *Science in Marketing*, New York: John Wiley and Sons, Inc., pp. 398-429.

Wadinambiaratchi, G. (1965). "Channels of Distribution in Developing Economies," *The Business Quarterly*, Winter Vol. 30, pp. 74-82.

SECTION II: MACRO INTERNATIONAL MARKETING ISSUES

Chapter 2

Overcoming Barriers to Trade in the Asia Pacific Region: Perspectives, Cooperative Agreements and Governmental Policy

Patricia Dunn Duffy

INTRODUCTION

The challenges facing business professionals in encouraging freer product exchange and trade in the Asia Pacific region are greater than simply a notion of "East meets West" or a quickly formulated business deal. There are differences in policy and perspective which can not only affect the outcome of business negotiations, but the entire viability of a transaction. Business professionals must understand and cope with a variety of internal and external factors in the form of trade barriers which can inhibit the free exchange of products across national borders.

In this chapter, we examine the more common barriers to trade, such as tariffs, duties, and non-tariff barriers (NTBs); and many of the policy arguments advanced by nations to protect their own domestic industries—often to the point of straining Asia Pacific relations. However, there needs to be greater emphasis on increasing economic cooperation and the freer movement of products across national boundaries. Also discussed is the growing emphasis upon and influence of cooperative agreements, and how bilateral and multilateral trade agreements are affecting both international relations and business growth in the Asia Pacific region. Only through increased trade and reducing barriers can distances be overcome which have separated both people and nations for centuries.

OVERVIEW OF BARRIERS TO TRADE

Trade is frequently viewed as one of the vital forces in today's global business market. Due to increased trade, "Pacific basin nations are growing more rapidly than any in the world," resulting in greater economic growth and cooperation in the Asia Pacific region (Gelman, 1984, p. 76). The benefits of trade include a freer flow of products and ideas, a greater cultural exchange, growth in standards of living, and improvements in economic welfare (Laffer and Miles, 1982; Robock, Simmonds and Zwick, 1977). Trade encourages better cooperation between both nations and global businesses. International product exchange also assists in promoting domestic welfare by reducing unemployment (Schwartz and Volgy, 1985). Trade creates and helps preserve jobs (Brown, 1986). According to a United States Presidential report to Congress in 1980, exports accounted "for one in every eight jobs in America's factories, and one in every four on America's farms" (Carter, 1980, p. 1).

Unfortunately, some people mistakenly view trade as a threat to domestic industries and many nations react to strong special-interest group pressure to inhibit foreign imports through enactment of various trade barriers. "World trade is a living organism whose whole health suffers if part of it is indisposed" (Curzon and Curzon, 1970, p. 1). It must be realized that trade is a two-way street. Trade consists of both imports and exports. Both importing and exporting firms and nations must realize that for effective trade to take place, there must be both give and take by the trading parties. It is unrealistic to expect one nation to permit importation of a foreign nation's products when that foreign country will not accept theirs. Only through improved inter-regional cooperation can firms enter foreign markets and retain significant foreign market share.

Global businesses must encourage freer trade rather than promoting protectionist methods, such as barriers to product exchange, in order to ensure long-term survival. As Robock, Simmonds, and

Zwick (1977, p. 121) clearly point out, "free trade maximizes world output." It is well established that nations "which have pursued liberal trade policies (Western democracies) have by and large experienced high rates of trade growth at higher absolute levels than those which have been more protectionist" (Curzon and Curzon, 1970, p. 5). Certain industries, and usually particular firms or government enterprises, may garnish certain economic benefits from protection through trade barriers. Unfortunately these benefits are temporary at best and serve only to postpone real problems for businesses. Additionally, while the protected industry enjoys its seemingly utopian position, overseas competitors are improving processes, increasing efficiencies, reducing costs and developing superior products. Thus, postponing the inevitable through protectionism only results in long-term disaster.

Those involved in global business must have an understanding of the various impediments to international product flow. The free flow of trade between nations is often inhibited by various forms of overt and covert trade barriers. General classes of overt barriers can include quantitative restrictions, tariff and non-tariff barriers (NTBs). Less obvious barriers can include legal and administrative impediments, product safety and quality standards; and social, cultural and political barriers to foreign product entry and acceptance. As Kotler (1986, p. 118) indicates, covert barriers to trade "may include discriminatory legal requirements, political favoritism, cartel agreements, social or cultural biases, unfriendly distribution channels, and refusals to cooperate."

Politics and Policy of Trade Barriers

Tariffs in the form of duties and taxes are imposed for a variety of political and economic reasons. They are a means to collect revenue, and restrict the importation of competing goods in an effort to protect domestic industries (Ethier, 1984). Trade barriers can also be imposed for political reasons. They can be a form of retaliation against the imposition of barriers aimed at exports by another nation, or to condemn political policies in the form of trade sanctions or complete prohibition against imports. An example is the banning of imports from the People's Republic of China to the United States until the mid-1970s, and current trade prohibitions between Vietnam and the U.S. However, protectionism is frequently the greatest political motivation behind increased use of trade barriers.

Protectionism is both a symptom and cause of economic difficulties in a nation. In Morrow's (1983, p. 68) classic essay on protectionism, he noted that protectionism can lead to world economic decline and possibly war.

> The famous [U.S.] Smoot-Hawley Tariff Act 1930 set up the highest tariff rate structure that the U.S. had ever had. One nation after another retaliated. The tariffs helped deepen the Great Depression worldwide and thus at least indirectly brought on World War II.

Protectionism, in effect, is a highly contagious political disease. It manifests itself in the form of political pressure on law-makers, anti-foreign public sentiment, and economic withdrawal. Imposing trade barriers during a time of political protectionist sentiment is like an alcoholic trying to cure a hang-over with a few more beers: a little hair of the dog that bit him. Morrow (1983) explains it best.

> When an economy gets sick, it wants to withdraw from the world. A protectionist psychosis sets in. The invalid retreats into the house and locks the doors and windows and then pulls the shades. Hypochondriac, jittery, paranoid, the economic system settles down to feed upon its own inadequacies. It sits in its slippers by the cold furnace and thinks about how well it used to make things, long ago. It disconsolately guzzles Old Smoot-Hawley, far into the night. Then it passes out. Another economy gone, as defunct as Mayan civilization.

The politics and policy associated with the increase in the use of NTBs as a means of domestic protection are highly complex. Often, they are imposed as a means of coercing foreign firms and governments into foreign direct investment and to force the exportation of *industries* rather than goods (Hollerman, 1984, p. 311). What this ultimately involves is the discouragement of foreign goods imports in favor of importation of capital and industries where the host (importing) nation receives the benefits of economic and capital improvement coupled with more domestic jobs created by these foreign-owned industries. To circumvent quota and tariff restrictions, foreign firms are forced to produce the goods in the host nation.

Consequences of Trade Distortion

Trade barriers distort the dynamics of international trade (Brown, 1986; Curzon and Curzon, 1970; Rieber, 1981). There have been unintended consequences which affected economic growth in third countries. "When each country independently introduces barriers without anticipating the reactions of others, however, the reduction in trade can be costly to all countries" (Robock, Simmonds and Zwick, 1977, p. 121). An example is television production and exports from Japan, South Korea and Taiwan. According to Wortzel and Wortzel (1981, p. 56), a 1976 orderly marketing agreement which restricted television imports into the United States from Japan resulted in a strong stimulus to television production and exports to the U.S. from South Korea and Taiwan. By constricting Japanese TV imports in an effort to protect domestic television manufacturers from foreign competition, the U.S. inadvertently stimulated competition from Taiwanese and South Korean manufacturers who shipped over one million TV sets to the U.S. in 1978. This illustrates the problem with some bilateral agreements which often simply substitute products from one country with the products from another.

This boom to Taiwanese and South Korean producers was, however, quite short-lived since "a subsequent orderly marketing agreement concluded with South Korea and Taiwan now limits the two countries to a total of 500,000 sets per year" (Wortzel and Wortzel, p. 56). This subsequent constraint on TV manufacturers in South Korea and Taiwan had devastating effects on those industries since the radical limitation on exports to the U.S. affected those nations' distribution infrastructures. As Wortzel and Wortzel (p. 56) point out, "the low prices at which they must sell require substantial volume in order to cover distribution costs."

Trade barriers have also distorted free competition by placing barriers to market entry which jeopardize entry by new firms (Hillman, 1982). An example is import quotas on clothing from Asia Pacific nations to the U.S. "The limited availability of quotas has made it difficult for new firms to enter an industry which had previously been extremely easy to enter but which offered low returns" (Wortzel and Wortzel, p. 56). Not only have new manufacturers been deterred in exporting countries, but free trade and competition has been inhibited whereby those firms possessing quota rights are given a monopoly on certain export markets. Another barrier to entry by new firms according to Hillman (1982, p. 1185) is that only existing firms reap the benefits while the total industry size declines, or at best remains stable, thereby discouraging newcomers.

Quotas have also distorted economies of scale and have caused product shifts toward more expensive items (OECD, 1984). "Quota limitations have also made it easier for the [newly industrialized country] producers to raise their prices and trade up to more expensive items" (Wortzel and Wortzel, p. 56). An example is the shift in exports by Japanese auto makers from inexpensive to luxury models. It is the consumer, and most frequently those with low incomes, who ultimately bears the cost of higher prices and who ends up subsidizing the profitability of foreign enterprises (OECD, 1984, p. 23).

Quotas result in a scarcity of many foreign goods which are perceived as better value for money due to superior quality as compared with domestic products. Since the product is in scarce supply, domestic retailers can charge a premium. Tagg (1984) believes that the extra costs added to quota goods result in a form of hidden tax to consumers and is greatly exemplified by the Japanese auto market in the U.S. Domestic enterprises may also unfairly benefit at the expense of the consumer (McGrath, 1983).

> When foreign cars are more expensive because supply falls short of demand, American auto makers can sell their own products at higher prices, too, because of the absence of price competition. According to the Congressional Budget Office, cars sold in [the U.S.], both foreign and domestic, cost an

> average of $1,200 more than they would without import quotas. (Tagg, 1984)

This distortion essentially backfires on the nation imposing the quota on foreign products. It makes a great deal of sense for a foreign manufacturer who is limited in the number of product units it may export to a given country to concentrate on higher-price, higher-profit items. The consumer who purchases those imported products must pay more, thereby reducing available discretionary income to purchase domestic products and with a concurrent and significant capital outflow to the foreign producer. Savvy producers use the increased capital to improve production efficiencies which make competition with them even more difficult. An example is the Japanese auto industry which has wisely invested foreign capital and profits in increasing production efficiency with which the stagnant U.S. auto industry has found it almost impossible to compete with without entering into joint ventures and encouraging further protectionist measures. As Nevin (1983, p. 89) wisely points out, protectionist measures "limit the options available to [domestic] consumers and reduce the economic pressure on domestic producers to become more competitive." Thus, both domestic consumers and producers are the real losers.

Protectionist measures inhibit economic growth by creating false job security, promoting declining industries, and through inefficient allocation of resources. Trade barriers result in a misallocation of resources away from more productive enterprises (Hollerman, 1984; Rieber, 1981; "Recent trends," 1985).

> [Protectionism] probably destroys more jobs than it "saves," and those that are saved carry a hefty price tag. The declining health of the US steel industry is a prime example of the failure of protectionism. Quotas, trigger-price mechanisms, higher tariffs, and anti-dumping suits have not made [the U.S.] steel industry more competitive. All those measures, however, have raised costs for steel-consuming companies who are now also calling for protection. The Federal Trade Commission estimates that the 1984 extension of steel quotas

> cost consumers $113,622 per job saved. (Brown, 1986, pp. 1079-80)

Laffer (1983) also indicates that protection of the U.S. steel industry has been "counterproductive and largely unanticipated." The impact of protectionism on domestic welfare is not unique to the U.S. the same principles apply to all nations in the Asia Pacific region.

Export Fluctuations, Import Substitution and Protectionist Policy

A frequent political justification for imposing import restraints is that they help to create greater stability in the domestic economy by reducing the uncertainty effects of export fluctuations. The standard argument is that by reducing the effects of variability in national export levels, greater domestic economic welfare is enhanced. To further fuel the protectionist fire in an attempt to reduce "import dependence" or competition from foreign firms in the domestic market, countries adopt a policy encouraging domestic production in the form of import substitution and to concentrate less effort in export manufacturing. However, empirical research using cross-national data, with significant emphasis on Asia Pacific nations, reveals that the opposite is true and such policies are counterproductive.

According to studies performed by Moran (1983) and Tan (1983) no statistical significance can be found to be associated between export instability and domestic growth. Rather, domestic growth is more likely to be a function of intrinsic factors such as natural resources, size, and level of development (Nugent and Yotopoulos, 1982). More importantly, domestic growth and economic welfare are greatly enhanced by an export oriented policy rather than through import substitution (Balassa, 1978; Heller and Porter, 1978; Tyler, 1981). As Tyler (1981, p. 129) keenly observes, "countries which neglect their export sectors through discriminatory economic policies are likely to have to settle for lower rates of economic growth as a result." Other researchers, such as Feder (1982), have reached basically the same conclusion. Export growth and economic growth are inextricably linked. In particular, trade and economic growth are vital partners (Kravis, 1970; Salvatore, 1983).

Legitimate Uses of Trade Barriers

Some import and export disincentives or prohibitions may serve legitimate national goals. Justifications may include public health, safety and welfare; national security, promotion of human rights goals, arms control, or to deter war or political aggression. Other rationale include protecting the environment, conserving scarce national resources, and safeguarding property of historical or national significance. Under extreme and limited circumstances, trade barriers can be used to alleviate critical shortages and reduce balance of payment deficits. However, differences in governmental policies and political philosophies greatly affect the relative "legitimacy" of trade restraints.

DIFFERENCES IN GOVERNMENTAL POLICY AND PERSPECTIVES

Differences Between Market and Nonmarket Economies

Differences between market and nonmarket economies can have profound effects on global business within the Asia Pacific region. During the past half-century, the impact and influence of nonmarket economies has grown in this region, particularly in Asia. Nations with centrally controlled nonmarket economies include the People's Republic of China (P.R.C.), Laos, Kampuchea (Cambodia), North Korea, Vietnam, the U.S.S.R., and to a much lesser extent, Mexico and Nicaragua.

One of the main facets of centrally planned economies is that government ownership predominates, rather than private ownership of business and trading interests. "Planned economies differ from market economies in that they are not responsive to the forces of the laws of supply and demand; they are not 'market' oriented" (Giffen, 1973, p. 62). Various councils and ministries control import and export policy, production decisions, and trade negotiations.

> The common pattern for the socialist [or nonmarket economy] countries, . . . is for the government, through a ministry of trade, to exercise complete control over exports and imports. Foreign traders deal with such a ministry rather than directly with the consumer or local vendor of the goods and services. With most economic activities owned by the governments and operated in accordance with national economic plans, foreign trade is also managed according to national plans rather than market-determined commercial opportunities. Political as well as economic considerations can be decisive. (Robock et al., 1977, p. 143)

As Kapoor (1974) indicates, one of the key problems many global business people face is an insufficient understanding of the importance of the role of government in nonmarket economy nations, and in particular, the political differences between governmental attitudes towards business concepts. According to Milosh (1973, p. 69), "in East-West trade, there are different parameters that have to be applied, not only in what to expect, but also in how to proceed."

The main political difference between free market and nonmarket economies is generally political ideology: communism vs. capitalism. The ideological gap between these two political viewpoints is wide and difficult to close. Strict Marxist dogma views capitalism as an evil where profit motive preys on the worker; whereas business-people in the "capitalist" world view profits as a means of stimulating social welfare through better pay, more capital, increased efficiency, and greater economic growth. Profits, improved market share, and greater trade essentially mean a better standard of living and increased social welfare to the capitalist. The communist sees just the opposite.

Communist Philosophy and Its Impact on Political Policy and Perspectives Towards Trade

As indicated earlier, there is a distinct dichotomy between the political attitudes of nonmarket (communist) and free-market (capitalist) economies. While many nations in the Asia Pacific region may have adopted political viewpoints somewhere in between these two extremes, such as non-Marxist socialism, the critical distinctions between communist and capitalist philosophies towards business and trade are important.

Those in the "free world" who desire to do business in the Asia Pacific region need to understand how the people who make business decisions in communist countries think in order to plan effective business strategies. Over half the people who live in the Asia Pacific region are under some form of communist government rule, with most people concentrated in the P.R.C., the U.S.S.R., and other Asian nations bordering the Pacific. Ideology plays a great part in business dealings in these nations.

Most people know that communism was basically founded on the philosophies of Karl Marx and Friedrick Engels, and the first major nation to use Marxism as a basis for political rule was the Soviet Union. Under strict Marxist doctrine, capitalism is the "evil empire" and capitalists are exploiters of the working class who must be overcome (Marx and Engels, 1888, repr. 1985, p. 95). According to strict Communist party foundations, which are still taken very seriously today in many communist nations, there must be the total abolition of privately owned businesses, private property, private industry, private enterprise, profits and capital (Marx and Engels, pp. 82-105).

According to strict doctrine, the capitalist tools of free trade, marketing and business practices, are not perceived favorably. Thus, "capitalists" who seek to enter into business negotiations with government representatives in nonmarket economy countries need to be aware that although trade representatives may appear somewhat receptive to a business deal, their fundamental teachings tell them that they must be extremely wary of capitalist advances. Regardless of how closely the government really adheres to strict Marxist policies, there are still fundamental communist precepts that must be, at least outwardly, adhered to.

Regardless of what free-market business executives may think, the vast majority of people they will deal with in communist countries are dedicated government representatives who are in their positions because of their adherence to communist party doctrine. The actions of these trade representatives are very closely monitored to ensure that any contracts they enter into on behalf of their government meet basic criteria. The basic criteria are that the main facets of the deal serve some primary benefit to their society and that national goals are furthered. This explains one of the basic motivations for countertrade and counter-purchase agreements in these nations.

Even though strict Marxist policies had to be modified because the communist nations found it difficult to be totally self-sufficient in an integrated world, today's nonmarket economy leaders are very dedicated to their cause. They will deal with capitalist businesses only to the extent they have to, however, close cooperation is considered to be betrayal (Lenin, 1983 repr., p. 127). Current reforms in China indicate that although some trade is desired, there will still be little tolerance of "bourgeois liberalism" and that strict party discipline must be observed (Doerner, 1987, p. 19).

The Growth of Business Ventures Between Communist and Free Market Nations

During the past decade, leading communist nations in the Asia Pacific region, namely the U.S.S.R. and the P.R.C., have come to realize that despite ideological differences, trade with capitalist nations is of vital importance for economic prosperity (Gorbachev, 1986). Such a move was unthinkable and deplorable twenty or forty years ago.

> The number of East-West ventures is growing. A recent example is the venture between North Korea and two Japanese trade companies, Asahi Corporation and Ryuko Trading Company, to set up a chain of 31 stores in North Korea to sell goods from Japan and the Communist bloc. (Perlmutter and Heenan, 1986, p. 136)

While the vast majority of cooperative trade agreements involving nonmarket economy (NME) nations will be between those with similar political views, there are still plenty of business opportunities available to free market enterprises. While many barriers to trade will continue based on political and ideological differences, in some cases they are not insurmountable provided differences in cultural perspectives are understood and form the basis for business decisions.

Problems in Applying Anti-dumping and Subsidy Standards to Non-market Economy Nations

Many free market economy nations have encountered difficulties in assessing anti-dumping and countervailing duties on imports from nonmarket economy countries. Anti-dumping duties are assessed by an importing country to counteract any unfair price advantage when a product is sold abroad or "dumped" at prices below their domestic or actual price. For example, selling radios which cost five dollars to produce at an import price of two dollars is considered "dumping." Simplistically, in free market economies, imposing an anti-dumping duty of three dollars would lessen any unfair advantage. Countervailing duties are imposed to counteract any unfair advantages which have been accrued where products have been heavily subsidized by government.

The problem of assessing countervailing and anti-dumping duties is quite complex due to the nature and structure of nonmarket economies. Essentially, the NME government subsidizes or controls production so that it is difficult to equate a fair value price for a NME-produced product with those produced in a free-market economy. While this problem has been recently addressed by the U.S. International Trade Commission, and other nations may resolve the issue as a result of further negotiations of those nations who are parties to the latest General Agreement on Tariffs and Trade (GATT) round in Uruguay, it still remains a current problem for many market economy nations.

Since actual cost of the goods from NME nations cannot be determined, constructive value must be used. Determination of constructive value is generally made by comparing the price of a like product sold in a third, market economy country. If no like product price or constructive value can be fairly determined, then many nations take the approach of taking the actual price of the product and adjusting the price to accommodate for a theoretical profit margin, such as 8 or 10%. Many free market nations, such as Australia, Canada, New Zealand, and the E.E.C. members, permit their authorities great latitude in determining equivalent or constructive prices from imported goods from nonmarket economy nations (Horlick and Shuman, 1984).

Overcoming Differences in Cultural Perspectives

Cultural differences between peoples can result in indirect, but significant barriers to trade and business cooperation (Hallen and Wiedersheim-Paul, 1980). What constitutes a nation's culture, or even various subcultures, is inherently complex. Culture generally encompasses "the whole set of social norms and responses that condition a population's behavior" (Robock et al., 1977, p. 309). Culture affects consumer buying habits and patterns, product promotion, business negotiations and a host of social responses.

According to Douglas and Dubois (1977), a nation's cultural environment affects business and marketing opportunities through its influence on social values, societal organization, rituals and mores, communications, life-styles, and all social interactions. Henry (1976) observes that consumer satisfaction is inextricably linked to cultural values. Culture also influences perceptions of foreigners and foreign products, prejudices, nationalistic policies, and to some degree, protectionism (Rierson, 1968; Schooler, 1965, 1971).*

Culture is a learned pattern of behavior which distinguishes one group of people from another. Culture is also inextricably linked with language and "each culture reflects in its language what is of value to the people" (Robock et al., 1977, p. 311). Thus, language is an important cultural variable which may result in indirect barriers to trade, product acceptance in the form of promotional message and product instructions, and business dealings.

Nonverbal patterns are also important. Nuances and nonverbal

---

*For a slightly different approach, see J.K. Johansson, S.P. Douglas, and I. Nonaka, "Assessing the impact of country of origin on product evaluations: a new methodological perspective," *Journal of Marketing Research*, vol. 12 (November 1985), pp. 388-96.

communication can affect Asia Pacific trade relations on a person-to-person basis (Graham and Sano, 1984; Hornik, 1980). As Hornik (p. 36) observes, difficulties encountered in intercultural communications is not only linked to understanding verbal messages, "but of understanding nonverbal signals that are generally coded so automatically within a single culture that we are quite unconscious of them."

A vital key to global business success in the Asia Pacific region is an understanding and appreciation of the culture of the people with whom businesses must deal with and the cultural environment in which business negotiations are conducted. Research has shown that one of the key elements for business success in the Asia Pacific region is a cultural understanding of the target market for imported products and the people who make business decisions. Graham (1981) indicates that one of the problems associated with Japanese and U.S. trade negotiations is "cultural myopia" which can, in effect, cause significant losses in business sales.

Cultural misunderstandings, ethnocentrism and cultural myopia can result in significant barriers to trade in the Asia Pacific region. Surveys of U.S. businessmen who have had experience in trading with the P.R.C. compiled by Brunner and Taoka (1977) reveal that business opportunities and negotiations are much more successful where there is greater cultural and political understanding. For example, just as Graham (1981) and Graham and Sano (1984) have observed with Japanese and American business negotiations, Americans desiring to do business in the P.R.C. must exhibit greater patience than is normally required in Western negotiations (Brunner and Taoka, 1977, p. 72).

LESSONS FROM THE JAPANESE AND AMERICANS

Trade difficulties between the Japanese and Americans pose significant lessons for business interactions within the Asia Pacific region. Two different cultures have had to adjust to differences in personal, social and business perspectives which have drastically affected trans-Pacific trade.

While marketing theory and practice may have had strong origins in the U.S., no one can deny that the Japanese have been effective in developing international marketing strategies which have propelled Japan into a world-class trading nation. Japan's successes in marketing, management, and production have frequently baffled outsiders. This has resulted in greater academic and business interest (Hayes, 1981; Lazer, Murata and Kosaka, 1985; Ozawa, 1982; Pascale and Athos, 1981; Suzuki, 1980; Vogel, 1979; Weiss, 1984).

> It is exceptionally difficult for a *gaijin*, a foreigner or outsider, to understand what is really going on in Japanese marketing because of a lack of understanding of the culture, language, and the historical perspective of Japanese business developments. (Lazer, Murata and Kosaka, 1985, p. 69)

Fortunately, the cultural gap is closing due to greater educational efforts and cultural awareness across the Pacific. Cross-cultural and product exchange, and sharing of ideas and technology has resulted in stronger ties between the two countries. However, despite many cultural bridges spanning the gulf between the two countries, a new wave of protectionist sentiment appears to be causing friction between the two nations.

> In the years following World War II, wise and sensitive leaders on both sides of the Pacific built a relationship between Japan and the United States that has been solidly grounded on feelings of respect and affection between the people of the two nations. That relationship is now seriously threatened by acrimonious disagreements over trade matters. (Neven, 1983, p. 95)

Nations in the Asia Pacific region can learn many lessons from the current discord between the Japanese and Americans. Despite decades of cultural exchange, understanding and progress, relations between the two nations and their peoples are strained due to trade barriers.

The Trade Deficit Myth

The current trade donnybrook between Japan and the U.S. illustrates how many misconceptions—and even outright errors—can greatly influence trade relations in the Asia Pacific region. Many American politicians and businesses are crying for more protectionism against foreign products, particularly Japanese, due to a growing U.S. trade deficit and in retaliation for real and perceived barriers to U.S. products in Japan. Despite trade deficits with other Asian and Pacific nations, Japan appears to be currently bearing the brunt of protectionist efforts. However, if protectionists in the U.S. succeed in attempting to rectify the current trade imbalance with Japan, there could be devastating effects for many Asian and Pacific trading nations. For example, in 1982, the U.S. had trade deficits with Japan, Canada, Hong Kong, Taiwan, the Philippines, Indonesia and South Korea. This is despite concluding several orderly marketing agreements and voluntary export restraints involving numerous "competing" products.

Curiously, the U.S. has not behaved consistently regarding balance of trade figures. For example, during 1982, the U.S. had trade surpluses with Australia, New Zealand, Mexico, Singapore, and the People's Republic of China. No mention is made of stabilizing the balance of trade with these nations. Further, despite a trade deficit in 1982 with Canada, the U.S. is currently making moves to establish a free trade zone between the two countries rather than encouraging greater protectionist measures as in the case of Japan.

Frequently, protectionists ignore facts and draw hasty conclusions (Moore, 1983; Schwartz and Volgy, 1985; Ueda, 1983). U.S. protectionists, for example, claim that the trade deficit and foreign imports have hurt the economy, U.S. industries, and have resulted in rampant unemployment. Table 2.1 illustrates the contrary, however. Despite a 426% increase in the trade deficit from 1970 to 1980, the value of U.S. exports has risen almost 400%, the gross domestic product (GDP) has increased 162%, and per capita GDP (GDPC) has increased 138%. From these figures it would appear that the trade deficit has *helped* the U.S. economy rather than injuring it.

Schwartz and Volgy (1985) contend that the facts and figures

TABLE 2.1. Changes in Trade and Economic Indicators from 1970-1980

Indicator	1970	1980	Percent increase
Exports ($US bil.)[a]	53.7	264.3	392
Trade Deficit ($US bil.)[a]	-54.3	-285.6	426
GDP ($US bil.)[a]	985.4	2586.4	162
GDP per capita ($US)[b]	4799	11416	138

SOURCES: [a]IMF International Financial Statistical Yearbook 1983

[b]UNCTAD Handbook of Int'l. Trade and Development Statistics

simply do not support claims of U.S. manufacturing and economic decline, and that the trade deficit argument for protectionism belongs in the same category as claims that the world is flat or in the center of the universe. Schwartz and Volgy indicate that U.S. manufacturing and productivity rose significantly during the 1970s and that the decline in the U.S. balance of trade was largely caused by the price of oil, rather than foreign goods.

> In sharp contrast, the U.S. balance of trade in manufacturing exhibited no slippage in these years. Starting with a favorable balance of $4 billion in 1970, the manufacturing trade balance remained unchanged at a positive $4 billion in 1979 and then rose during the 1980 recession to $18 billion. (Schwartz and Volgy, p. 105)

Moore (1983) and Ueda (1983) point out that the technique used to determine balance of trade figures can greatly affect their validity. Hence, politicians can select the means of measurement which best suits their point of view.

Schwartz and Volgy (1985) also explode the myth that the U.S. economy was depressed during the 1970s, as indicated by unemployment figures. They indicate that the growth in new jobs in the U.S. exceeded that of Japan with a 26% increase in the number of U.S. jobs during the 1970s (p. 100). Further, they contend, that rises in unemployment during the period of 1965 to 1980 was primarily the result of intrinsic rather than extrinsic factors, primarily the composition of the work force.

> Several factors brought more than 55 million new workers into the American labor market in the 15 years between 1965 and 1980: postwar baby boomers reached adulthood, single-person and single-parent families grew dramatically, and women changed their attitudes toward employment. This tidal wave of Americans looking for work caused a net surge of 30 million, a 40% increase between 1965 and 1980. (p. 101)

Another intrinsic factor which may also be contributing to any real decline in trade competitiveness of U.S. industries is domestic taxation policies (Low, 1982; Nevin, 1983).

Gelman (1984) reports that the focus on the balance of trade with Japan is myopic and ignores the fact that Japanese foreign direct investment in the U.S. has nearly equalled the capital outflow caused by U.S. purchases of Japanese goods. In other words, most of the hundreds of billions of U.S. dollars that flowed out of the U.S. due to purchases of Japanese products has come back in the form of Japanese investment. According to Gelman, that money has been used to actually improve the U.S. economy through financing the budget deficit and capital improvements in manufacturing facilities; has helped keep inflation down due to the lower cost of foreign goods resulting in a better standard of living for Americans; and has actually resulted in the creation of more jobs than those "lost" due to any trade deficit or competition from Japanese products. Samuelson (1984) also contends that foreign investment has actually resulted in creating far more jobs than those lost through imports. He also notes that, "historically, there is little connection between trade deficits and jobs."

HOW BUSINESS PROFESSIONALS CAN HELP OVERCOME ASIA PACIFIC TRADE BARRIERS

Cultural Understanding

As previously indicated, cultural differences in the form of bias or misunderstandings can be a significant barrier to trade. As a general rule, nations do not trade, businesses do (Leighton, 1970). And businesses and governments are run by people. People make the decisions. Just as people make conscious or subconscious decisions to impede trade, they can also be encouraged to promote product exchange.

As Graham (1981), Graham and Herberger (1983), and Graham and Sano (1984) have clearly indicated, one of the greatest impediments to business dealings between Japanese and United States business and negotiators appears to be cultural bias, cultural myopia, and cross-cultural tensions. These difficulties are not unique to Japanese and American cultural perspectives, but are true of any cross-cultural business relationships and enterprises in the Asia Pacific region. "Business negotiations across cultures are difficult undertakings, particularly in such a fast-changing world. Much can go wrong besides the economics of the deal" (Graham and Sano, p. 112).

This is one barrier to trade that is surmountable and more easily within the grasp of business executives and negotiators within the Asia Pacific region. Not only are geographical, linguistic, and social differences important, people involved in global business must also understand the political perspectives of the target country and its people (Lee, 1966). Thus, businesspeople need to learn as much about the target foreign market and its people *before* entering into preliminary trade negotiations. This may involve additional behavior modification on the part of a potential exporter or importer so as not to inadvertently offend a foreign trade counterpart and thereby jeopardize trade negotiations.

Cooperative Ventures and the New Asia Pacific Industrial Revolution

Cooperative business ventures between firms in the Asia Pacific region are growing at a rapid rate, resulting in a new Asia Pacific industrial revolution. Many people are aware of the joint ventures between major automobile manufacturers in South Korea, Japan and the U.S. This is only the tip of the iceberg. A new wave of cooperation rather than competition is strengthening businesses both in the East and the West. The key phrase epitomizing the new Asia Pacific industrial revolution is: "if you can't beat them, join them." As Rosenberg and Van West (1984) contend, collaborative and cooperative business and marketing strategies can greatly assist domestic firms since they spend less time "fighting" competitors and concentrate their efforts in wooing customers.

Not only are cooperative business agreements, such as joint ventures, improving relations within the Asia Pacific region, all parties appear to be reaping great benefits through improved economies of scale, efficiency, and technological growth. While there have been some criticisms of these new alliances (Reich and Mankin, 1986), the benefits appear to outweigh any of the protectionist and "buy domestic" arguments.

Cooperative arrangements also are a great means of circumventing the deleterious effects of trade barriers. For example,

> to avert rising protectionist sentiment, Japanese companies are setting up plants in the United States, either as joint ventures or on their own; to obtain high-quality, low-cost products and components, U.S. companies are making joint venture agreements with Japanese companies. (Reich and Mankin, 1986, p. 78)

This is only the beginning. There is a great deal that business professionals can do to increase trade in the Asia Pacific region. Cooperation rather than competition may be a more appropriate means of stimulating trade (Ayal, 1981, p. 96). People, firms and governments can take a more active role in reducing both overt and covert barriers to freer product exchange. Businesses can promote education efforts to inform the public and politicians about the truths surrounding protectionist myths and the adverse affects of trade barriers.

People and businesses can take an active role in lobbying for increased cooperative trade agreements and reduced barriers. Firms and nations in the Asia Pacific region may wish to expand the scope of free trade area agreements to encompass the vast majority of Asia Pacific nations, such as the creation of an "Asia Pacific Free Trade Area." Another possibility could be an expansion of the Association of South East Asian Nations (A.S.E.A.N.) to include the "Association of Asian and Pacific Nations." As the classic saying goes: "united we stand, divided we fall."

REFERENCES

Ayal, I. (1981). International Product Life Cycle: A Reassessment and Product Policy Implications. *Journal of Marketing*, 45, 91-6.

Balassa, B. (1978). Exports and Economic Growth: Further Evidence. *Journal of Development Economics*, 5, 181-9.

Brown, C.J. (1986). U.S. Competitiveness in World Markets. *Journal of International Law and Politics*, 18, 1075-86.

Brunner, J.A. & Taoka, G.M. (1977). Marketing and Negotiating in the People's Republic of China: Perceptions of American Businessmen Who Attended the 1975 Canton Fair. *Journal of International Business Studies*, pp. 69-82.

Carter, J. (1980). *Report of the President on Export Promotion Functions and Potential Export Disincentives*. Washington D.C.: The White House.

Curzon, G. & Curzon, V. (1970). *Hidden Barriers to International Trade*. London: Trade Policy Research Center.

Doerner, W.R. (1987). Asia/Pacific (China): Putting Limits on Thought. *Time*, 23 February.

Douglas, S. & Dubois, B. (1977). Looking at the Cultural Environment for International Marketing Opportunities. *Columbia Journal of World Business*, 12, 102-118,

Ethier, W.J. (1984). Protection and Real Incomes Once Again. *Quarterly Journal of Economics*, 99, 193-200.

Feder, G. (1982). On Exports and Economic Growth. *Journal of Development Economics*, 12, 59-73.

Gelman, E. (1984). A Marriage of Convenience: Trade Between Japan and the United States Suits the Needs of Both Countries. *Newsweek*, 5 November, pp. 75-6.

Giffen, J.H. (1973). Developing a Market Program for the U.S.S.R. *Columbia Journal of World Business*, 8,61-8.

Graham, J.L. (1981). A Hidden Cause of America's Trade Deficit with Japan. *Columbia Journal of World Business*, 16.

______ & Herberger, Jr, R.A. (1983). Negotiators Abroad—Don't Shoot from the Hip. *Harvard Business Review*, July-August, 160-8.

______ & Sano, Y. (1984). *Smart Bargaining*. Cambridge, MA: Ballinger Publishing Co.

Gorbachev, M.S. (1986). Remarks on US-USSR Trade. *Harvard Business Review*, 64, May-June, 55-8.

Hallen, L. & Wiedersheim-Paul F. (1980) Psychic Distance and Buyer-Seller Interaction. *Centre for International Business Studies* (Uppsala, Sweden), 9, 308-24.

Hayes, R.H. (1981). Why Japanese Factories Work. *Harvard Business Review*, 59, July-August, 56-66.
Heller, P.S. & Porter, R.C. (1978). Exports and Growth: an Empirical Re-investigation. *Journal of Development Economics*, 5, 191-3.
Henry, W.A. (1976). Cultural Values Do Correlate with Consumer Discontent. *Journal of Marketing Research*, 13, 121-7.
Hillman, A.L. (1982). Declining Industries and Political-Support Protectionist Motives. *American Economic Review*, 5, 1180-7.
Hollerman, L. (1984) Japan's Direct Investment in California and the new Protectionism. *Journal of World Trade Law*, July-August, 309-19.
Horlick, G.N. & Shuman, S.S. (1984). Nonmarket Economy Trade and U.S. Antidumping/Countervailing Duty Laws. *International Lawyer*, 18, 814-7.
Hornik, J. (1980). Comparative Evaluation of International vs. National Advertising Strategies. *Columbia Journal of World Business*, 15,36-45.
Johansson, J.K., Douglas, S.P. & Nonaka I. (1985). Assessing the Impact of Country of Origin on Product Evaluations: A New Methodological Perspective. *Journal of Marketing Research*, 12, 388-96.
Kapoor, A. (1974). MNC Negotiations: Characteristics and Planning Implications. *Columbia Journal of World Business*, 9, 121-30.
Kotler, P. (1986). Megamarketing. *Harvard Business Review*, 64, 117-24.
Kravis, I.B. (1970). Trade as a Handmaiden of Growth: Similarities Between the Nineteenth and Twentieth Centuries. *Economic Journal*, 80, 850-72.
Laffer, A.B. (1983). Limits on Trade Won't Spell Relief. *Los Angeles Times*, 11 January, sec.I, p. 3.
______ & Miles, M.A. (1982). *International Economics in an Integrated World*. Palo Alto, California: Scott Foresman and Co.
Lazer, W., Murata, S. & Kosaka, H. (1985). Japanese Marketing: Towards a Better Understanding. *Journal of Marketing*, 49, 69-81.
Lee, J.A. (1966). Cultural Analysis in Overseas Operations. *Harvard Business Review*, 44, March-April, pp. 106-14.
Leighton, D.S.R. (1970). The Internationalization of American Business—the Third Industrial Revolution. *Journal of Marketing*, 34, 3-6.
Lenin, V.I. (1983 reprint). *Selected Works*. CT: Hyperion Press.
Low, P. (1982). The Definition of "Export Subsidies" in GATT. *Journal of World Trade Law*, 16, 375-90.
Marx, K. & Engels, F. (1888). *The Communist Manifesto*. (1985 reprint), Middlesex, England: Penguin Books Ltd.
McGrath, A. (1983). Import Quotas: The Honda Dealer's Best Friend. *Forbes*, 5 December, pp. 43-4.
Milosh, E.J. (1973). Imaginative Marketing in Eastern Europe. *Columbia Journal of World Business*, 8, 69-72.
Moore, G.H. (1983). Will the "Real" Trade Balance Please Stand Up? *Journal of International Business Studies*, Spring-Summer, 155-9.

Moran, C. (1983). Export Fluctuations and Economic Growth. *Journal of Development Economics*, 12, 195-218.

Morrow, L. (1983). The Protectionist Temptation. *Time*, 10 January, p. 68.

Nevin, J.J. (1983). Doorstop for Free Trade. *Harvard Business Review*, March-April, 88-95.

Nugent, J.B. & Yotopoulos, P.A. (1982). Morphology of Growth: The Effects of Country Size, Structural Characteristics and Linkages. *Journal of Development Economics*, 10, 279-95.

OECD (1984). The Consumer: A Force Against Protectionism. *OECD Observer*, July, 22-5.

Ozawa, T. (1982). *Multinationalism, Japanese style*. Princeton University Press, Princeton, New Jersey.

Pascale, R.T. & Athos, A.G. (1981). *The Art of Japanese Management*. NY: Warner Books.

Perlmutter, H.V. & Heenan, D.A. (1986). Thinking Ahead: Cooperate to Compete Globally. *Harvard Business Review*, 64, March-April, 136-51.

Reich, R.B. & Mankin, E.D. (1986). Joint Ventures with Japan Give Away Our Future. *Harvard Business Review*, 64, March-April, 78-86.

Rieber, W.J. (1981). Tariffs as a Means of Altering Trade Patterns. *American Economic Review*, 5, 1098-9.

Rierson, C. (1968). Attitude Changes Toward Foreign Products. *Journal of Marketing Research*, 4, 385-90.

Robock, H., Simmonds, K. and Zwick, J. (1977). *International Business and Multinational Enterprises*. Homewood, IL: Richard D.Irwin, Inc.

Rosenberg, L.J. & Van West, J.H. (1984). The Collaborative Approach to Marketing. *Business Horizons*, November-December, pp. 29-35.

Salvatore, D. (1983). A Simultaneous Equations Model of Trade and Development with Dynamic Policy Simulations. *Kyklos*, 36, 66-90.

Samuelson, R.J. (1984). The Global Money Game. *Newsweek*, 5 November, p. 82.

Schooler, R. (1971). Bias Phenomena Attendant to the Marketing of Foreign Goods in the U.S. *Journal of International Business Studies*, pp. 71-80.

______ (1965). Product Bias in the Central American Common Market. *Journal of Marketing Research*, 2, 394-7.

Schwartz, J.E. & Volgy, T.J. (1985). The Myth of America's Economic Decline. *Harvard Business Review*, 63, September-October, 98-107.

Suzuki, Y. (1980). The Strategy and Structure of Top 100 Japanese Industrial Enterprises 1950-1970. *Strategic Management Journal*, 1, 265-91.

Tagg, J. (1984). We're Paying for the Edsel of Trade Policy, *Los Angeles Times*, 5 March.

Tan, G. (1983). Export Instability, Export Growth, and GDP Growth. *Journal of Development Economics*, 12, 219-27.

Tyler, W.G. (1981). Growth and Export Expansion in Developing Countries. *Journal of Development Economics*, 9, 121-30.

Ueda, K. (1983). Trade Balance Adjustment with Imported Intermediate Goods: The Japanese Case. *Review of Economics and Statistics*, November, 626-32.
Vogel, E.F. (1979). *Japan as Number One: Lessons for America*. NY: Harper and Row.
Weiss, A. (1984). Simple Truths of Japanese Manufacturing. *Harvard Business Review*, 62, July-August, 119-25.
Wortzel L.H. & Wortzel, H.V. (1981). Export Marketing Strategies for NIC and LDC-Based Firms. *Columbia Journal of World Business*, 16, 51-6.
Recent Trends in New Zealand's International Competitiveness (1985). *Reserve Bank Bulletin*, 48, 565-71.

Chapter 3

The Process of Technology Transfer Through Joint Ventures: International Marketing Implications

Syed Akmal Hyder
Pervez N. Ghauri

Joint ventures (JVs) have become more and more important in the recent decades both as technology-transfer projects in less developed countries (LDCs) demanded by host governments and as a mode of foreign investment by multinational firms (MNCs). The present subject matter has received a considerable degree of importance by the researchers and there are a number of studies available on joint venture relationships between MNCs and LDCs (see for example, Vernon, 1972, Ahn, 1980, Walmsley, 1982 and Beamish, 1985). The relationship is, however, still facing great problems and the parties involved face difficulties in resolving their differences. The resources brought in by different partners, their organizations, their social and cultural backgrounds and the environment in which JVs are to operate are more or less unknown to each other. Thus, management of this relationship cannot succeed unless these differences are perceived and overcome. The available literature put forward several such management problems to discern and analyze the immediate conflict (see, for example, Wright, 1979 and Simiar, 1983). Most of these works do not, however, investigate and explain ways of solving current problems but rather concentrate on suggesting some measures to be taken prior to the formation of a JV. In other words, several studies are concerned with conflicts in

JVs but hardly any study focuses on the resolution of these conflicts.

There are thus still several questions left unresolved in the field, such as why, in fact, partners need to form a JV, what resources they contribute over time, why conflicts arise and how these are settled. It is almost understood that if a partner could do the job alone, he would probably not agree to share ownership and management responsibility with the other.

As mentioned earlier one reason for increasing interest in JVs is that it is a very attractive form of foreign investment where the MNCs can overcome the uncertainties and share the risk with others, the rate of failure in JVs is however, very high (Simiar, 1983 and Beamish, 1985). One of the reasons for this high rate of failure is that MNCs are most probably not aware of the dynamism in the JV relationships—that cooperation and conflict can change over time. A JV is a dynamic process and the problem and advantages related to it should be treated in a process perspective. As commented by Harrigan (1984, p. 8) joint ventures are formed within dynamic environments and therefore it is necessary to study a JV more intensively as it develops.

The present study has a longitudinal approach so that it can investigate reasons and consequences of partners' interactions in order to describe the development of the JV relationship over time. This work will be helpful to practitioners to understand day-to-day problems in JVs. Researchers in the field will be benefited by its view of considering a JV as a historical process. A longitudinal case study of a JV between a Swedish firm and an Indian partner is presented which covers a period of 15 years.

In this study the following questions are addressed:

- What stages does a JV, between a Western firm and a local partner from a LDC, go through?
- How is control exercised in a JV?
- How and when do conflicts arise between the partners and how are these resolved?

BACKGROUND

Using a sample of 166 American domestic JVs, Pfeffer and Nowak (1976) assumed that firms would resort to JV to manage interorganizational interdependence. Unlike Pfeffer and Nowak, Aiken and Hage (1968) found that the formation of joint activities was a reason to create interdependence in order to satisfy different resource needs. We argue that JVs are formed to manage interorganizational interdependence as well as to create interdependence. Edström (1975) also had a similar view. In their extensive and detailed separate studies, Tomlinson (1970) and Janger (1980) observed that local government pressure, local facilities and spreading risks were the major reasons for foreign firms to form a JV. Ahn (1980) discovered that local firms were attracted by foreign capital, technology, training, management and marketing know-how.

In general, the reason to form a JV is to gain complementary resources from each other, and an important characteristic of resource exchange is its dynamic nature (see Bivens and Lovell, 1966; Otterbeck, 1979; Berg and Friedman, 1980 and Connolly, 1984). Franko (1971) criticized the view of partners' contributions as a static bundle of skills or inputs; he argued that they were subject to change. Killing (1982 and 1983), who studied JV success, also considered the change in the need for resources from one another due to the presence of continuous learning. Koot (1986) and Harrigan (1986) also observed the divergence of business interests among the partners over time.

Tomlinson (1970) and Friedmann and Beguin (1971) were in agreement that foreign firms who desired to have control required majority ownership in the JV. Gullander (1976, p. 110) argued that control of critical decisions is not always determined by management control. He said, "For example, a parent who can control the supply of raw materials to the JV possesses power that could reach in decision areas other than those directly related to his immediate area of control, i.e., raw materials." Abdul (1979), Ahn (1980), Otterbeck (1979 and 1980) and Dunning (1978) also found critical resources as the sources of control. We argue that mere ownership is insufficient in exercising control: rather, supply of critical resources are associated with effective control. It is, therefore, important that both types of control are recognized in a JV study.

"Lack of goal congruence" has been cited by many authors as a major reason for JV conflict (see Reynolds, 1984; Edström, 1975 and Simiar, 1983). Barlew (1984, p. 50) commented in this connection, "The partners' business objectives may differ so radically that agreement on how to operate, fund, and benefit from the venture may be difficult to reach." Beamish and Lane (1982) and Raveed and Renforth (1983) recognized cultural differences as a source of problems in JVs, especially in the developing countries. We understand that conflicts are common in a JV. It is necessary to try to solve those problems in time before they go out of control. Unfortunately, most studies are not concerned with resolution of problems, and the reasons of JV failure remain more or less unexplored.

METHODOLOGY

As already mentioned, in contrast to most work on JV, this study seeks to examine the relationship over time. In this longitudinal study, all major happenings over time are duly investigated, discussed and finally analyzed. Due to the extensive nature of this work, it was possible to follow the exchange of resources, the origin of conflicts, and the gradual development of the relationship.

The data were collected via face-to-face interviews, telephone interviews, correspondence and printed materials.[1] Face-to-face interviews, which were conducted between 1982-1985 provided valuable information. These interviews were semi-structured and the interviewees were free to talk. The length of each interview varied from two to four hours depending on the depth and requirement of the discussions. The local partner, the general manager of the JV and the foreign partner's main representative were interviewed. Telephone interviews served two purposes, to arrange face-to-face meetings at the beginning and to collect complementary information from time to time. A questionnaire was prepared on the basis of the initial research problems after a review of previous JV studies, which are made for a similar purpose.

For case analysis, two matrices are used to discuss expected and actual contributions of the partners in connection with five resource variables. "Expected contribution" derives from the notion that each participant expects a requital during the period of collaboration. "Actual contribution" implies what the partners really contribute over time. In describing actual contribution, three parties, i.e., the local partner, the foreign partner and the JV itself are recognized as the latter gets institutionalized over time. To differentiate between formal control and control through resources, two more matrices are used. In the first matrix five determinants of formal control are considered, which are equity, board of directors, chairman, GM and plant manager. In the later case the five resource variables are applied.

The model used for the analysis consists of partners' interests, exchange of resources, control, cooperation and performance (Fig. 3.1). A common topic of discussion in most of the JV studies comprised the motives for creating such forms of cooperation. These motives or interests to form JVs are mainly a function of the partners' capability, overall position and critical mutual requirement. Partners' interests affect exchange of resources because they will

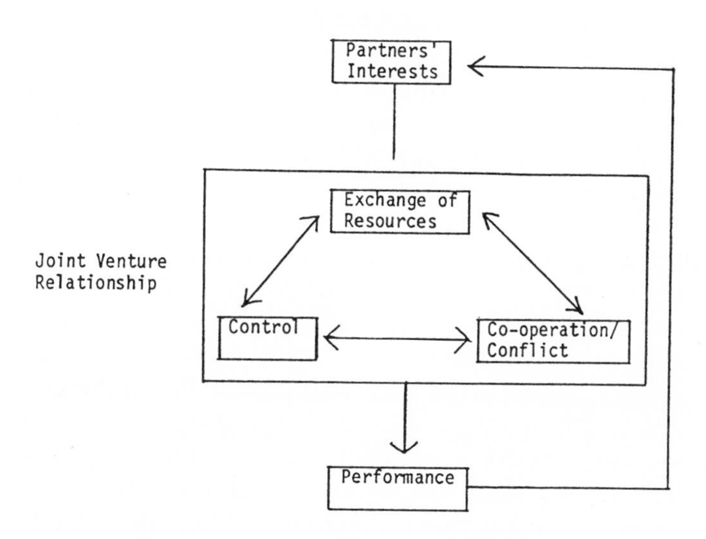

FIGURE 3.1. The Structure of the Model

decide how much of the available resources should be contributed to the JV. The control is a vital issue; does a partner need substantial control or will he be satisfied with no control? When a partner thinks that he needs the JV, he will try to cooperate but if his interest is not fulfilled through the JV, the opposite will happen.

In the model, exchange of resources affects control and cooperation/conflict. Pfeffer (1981, p. 68) identifies interdependence as the first condition for the use of power, thus mutual interdependence is established in a JV. Exchange of resources thus gives the partners control in return for their valuable contribution in the relationship. The exchange will bring cooperation if the contributions are made as anticipated. When a partner fails to provide an expected critical resource, conflict is more likely. Many authors have argued that conflict is embedded in the interdependence relationship (see Ghauri, 1983; Håkansson, 1986; Pfeffer, 1978 and 1981; Pfeffer and Salancik, 1978 and Schmidt and Kochan, 1972).

Control affects exchange of resources and cooperation/conflict. If a partner has formal control or access to a critical resource, he may refuse to supply the resource to take advantage. Control in a JV may also provide access to more resources. However, there are authors who maintain that control leads to conflict.[2]

At the same time cooperation/conflict has certain impact on exchange of resources and control. If partners are cooperative, they will try to give support with critical resources, even when it is difficult. Conflict between the partners may complicate the exchange of resources. When a partner can trust the other's competence, he will not usually hesitate to give control to the capable partner.

Every decision which the partners take may have some link with, or result in, cooperation/conflict, but may not necessarily be visible all the time. At the beginning, partners need to be highly cooperative, otherwise no agreement on sharing equity and responsibility can be signed. Conflict is very common in JVs because partners have different interests and have different backgrounds and values. In a JV, one common reason for conflict will be the differences between formal control and control through resources.

The reasons for entering a JV are not usually the same for both partners. The contributions made or contemplated by the partners pertain to their expectations. Ahn (1980, p. 197) describes that local partners are interested in receiving the following services from the foreign partners in addition to their equity capital contributions:

- credits or access to sources of finance;
- management know-how;
- technological know-how;
- marketing know-how and marketing facilities.

Each partner has to make some contribution which is significant to the relationship. Partners have extensive needs, and consequently, are not satisfied with short term contract or agreement and insist on a more stable relationship. Walmsley (1982, p. 2) observed the depth of the JV relationship, and found a deliberate alliance of resources between two independent organizations. The resources of the firms can be divided into five categories: input material, capital, manpower, technology and market. Input material has been described as one of the motivating factors for foreign firms to enter JVs in developing countries (Ahn, 1980, p. 196). The sources of capital to a JV are equity from both partners, in cash or kind, loans from partners, trade credits and loans from capital markets, local or abroad. In every organization, there is a need for manpower to carry out the operational activities according to the programs reached.

As partners put in capital, administrative efforts and reputation, they want to safeguard their interests through the exercise of control over the discretion and efficiency of the operations. Control has two facets, i.e., formal control and control through resources. Larsson (1985, p. 80) in a study of "power" in the transnational enterprise found that analysis of organizational power requires consideration of both formalized and nonformalized aspects that he called authority and influence, respectively. According to him, authority is dependent on ownership while influence will emerge from the exchange of resources.

Performance may be measured in relation to partners' interests and by common criteria of effectiveness. Performance in relation to partners' interests is discussed whenever partners have mentioned other goals to be achieved by the JV than just the JV success. To measure effectiveness, five criteria have been adopted from the works of Campbell (1977) and Steers (1980). These are profit, growth, adaptation, joint participation in the activities and survival. Profit means the amount of revenue from sales left after all costs and obligations are met. Growth implies a comparison of an organization's present with its past state. Adaptation refers to the ability of an organization to change its standard operating procedures in response to environmental change. Joint participation means that each partner has to perform some organizational function. Survival is the ability of an organization to exist without any basic change in structure and goals as set at the beginning.

THE CASE

The foreign partner Ericsson is one of the few giant Swedish multinationals having a leading position in the telecommunications industry of the world. Its total net sales amounted to 15.2 billion SEK (U.S. $1 = 5.98 SEK) during the first half of 1987. Technology transfer is an essential activity of Ericsson. It pursues no rigid policy regarding the form of investment. It has a large number of wholly-owned subsidiaries and both majority and minority JVs. The local partner is a well-established businessman in India and has been in the business for more than 30 years. His total investment in different businesses approach 5 million Rs (U.S. $1 = 14 Rs).

Ericsson had been working in the Indian market with a wholly-owned subsidiary for 10-15 years before the JV was formed. This subsidiary was basically a trading house and had difficulty in competing with the other foreign companies, which had local production facilities. The meeting between Ericsson's main representative and the local partner was a matter of chance. Personal liking played a major role in Ericsson's choice of the local associate. Ericsson's representative, who met his counterpart for the first time at the Swedish Embassy in India, found him charming. The local partner was approached by Ericsson and accepted the offer to set up a JV in January 1971.

Partners' Interests and Initial Expectations

JV formation was a necessity for Ericsson which could not otherwise engage in local production. From the beginning, Ericsson sought a capable local partner, whose contribution would complement its activities in India. Although the local partner played a passive role in the formation of the JV, he had definite objectives from the collaboration.

At the initial stage, a significant proportion of the input materials were expected to be obtained from external sources, not necessarily from Sweden.[3] The expected contributions of the partners and the criticalness of the resources are shown in Tables 3.1 and 3.2. Since the required input material for the JV were of an older type, Ericsson was not in a position to supply many of them from its own source. Ericsson was expected to provide information about the foreign suppliers and negotiate with them on behalf of the JV until the latter could import its own material. Most of the input material was, however, available in the local market but it seemed to take some time for the JV to know its needs and establish useful contacts with the local suppliers. As local suppliers were mainly foreign firms or big local companies, Ericsson was expected to play a vital role in this context.

TABLE 3.1. Expected Contribution of the Local Partner

Resource	Expected Contribution	Criticalness
Input Material	-	-
Capital	Making equity ownership by the payment of cash.	Very critical for the operational activity.
Manpower	Responsibility for the major recruitment.	Not critical as manpower abundant in the host country.
Technology	-	-
Market	Frequent meetings with the government buyers.	Very critical as legitimacy is a must.

Ericsson was not required to invest any cash during the time of JV formation, since its assets from Ericsson Telephone Sales Corporation sufficed to make up the share capital. Ericsson obtained major ownership while the local partner became the minority shareholder by making a cash investment. It was expected that the local partner would subsequently raise his share capital to comply with the government regulation. Ericsson was not able to invest any cash as the value of its assets already exceeded the amount of its share capital. Therefore, the cash contribution by the local partner was considered very critical for production, maintenance and development in the JV.

Extensive training of the local people both in Sweden and in India was planned. It was also expected that Ericsson would provide technological assistance to the JV in manufacturing and marketing complex products through participation in the local tenders. The permanent employment of expatriates in the JV was never discussed. As skilled and semi-skilled people were available in India, it was expected that the local partner would be able to recruit efficient manpower from the local market.

During the establishment of the JV, the local partner had no technical competence in the field and was wholly dependent on Ericsson

TABLE 3.2. Expected Contribution of the Foreign Partner (Ericsson)

Resource	Expected Contribution	Criticalness
Input Material	Procurement of all input materials at the beginning.	Critical for the assessment of the needs and making contact with the suppliers.
Capital	No cash investment	Not applicable
Manpower	Temporary arrangement of expatriates from time to time.	Critical in connection with start up and training.
Technology	Total transfer of technology and forwarding technical information.	Very critical to start and maintain the operating activity at the initial stage.
Market	Meeting buyers in connection to trading business.	Critical in supplying information to the buyers.

where technology was concerned. It was decided by both sides that the full technological responsibility would be handed over to the local people as early as possible. Since the total transfer had to be completed in a short time, Ericsson's role in this resource area was presumably very crucial.

It was a prime expectation that Ericsson's marketing experience from the wholly-owned subsidiary and knowledge of the products would be of great significance in respect to marketing. The local partner was a stranger in this field but his experience was greatly valued because the buyers were all government authorities, and he was well experienced in marketing other products all over India.

Exchange of Resources

Input Material

The local partner lacked knowledge of the requirements and also the sources of supply. Ericsson had both information and legitimacy in the eyes of the prospective suppliers. In the host country and abroad, Ericsson had the exclusive responsibility for the procurement of input material in the first two to three years. The JV gradually learned to handle this task alone while Ericsson's contribution remained important over the years to facilitate import of new items and supply some useful input material from its own source. There was one major problem to importing input material. The local government refused to grant import permits for the components which were available in the local market. These components, however, did not reach the requisite standard. In such a situation the JV could neither buy the indigenous products nor import them from abroad.

Ericsson was not able to supply most of the input material due to the age of the technology, the JV thus procured them from other sources. Ericsson supplied comparatively modern components, which were mostly related to the newly adopted products. The JV was free to buy from anywhere and from any company, which it found most suitable in respect to price and quality.

Capital

At the time of the JV formation, the initial shareholding of Ericsson was 74% while the rest was in the possession of the local partner. During the next seven years, the share capital ratio was progressively increased in favor of the local partner to comply with the government regulations. In 1978, the local shareholding reached 60.4% while Ericsson had 39.6%. This share structure remained constant at the end of 1985. In 1971, the total paid up capital was 1m Rs. which was raised to 1.9m Rs. in 1983. Ericsson was allocated 7,400 shares, of nominal value 100 Rs. each on the basis of its assets during the time of the JV formation. These assets were valued by the company's auditors. After the conversion of Ericsson's assets into shares, it invested no further capital.

For the Sahidabad project (second factory of the JV which was set up later on), the JV had been granted a loan by a local bank amounting to Rs. 0.69M, of which Rs. 0.27M had been received in 1980 and the rest during the first half of 1981. Besides this, the JV had also obtained credit facilities from American Express International Banking Corporation and the Bank of Tokyo Limited, Calcutta, during 1979-1980. Both the loans were secured but the reputation and influence of the local partner were very helpful. Ericsson was not known to play any role in this context.

Manpower

Recruitment was a primary responsibility of the local partner, since he had information and a close link with the local labor market. As the local partner was preoccupied with other business and the JV became institutionalized, the task of recruitment was gradually passed on to the general manager (GM), who was the chief executive of the company. However, the local partner was always consulted if some important appointments were to be made. The JV had only local employees, whose total number was around 100 in 1985.

The task of the GM was very difficult in the JV. He needed to be friendly with the local partner because he had day-to-day contact with him but at the same time had to assure Ericsson that his neutral stand was not controversial. There were altogether four top ranking executives including the GM in the JV at the end of 1985. Each was qualified and experienced and none was related to any of the directors. There were no expatriates permanently employed in the JV but Swedish personnel from Ericsson, whose work was mostly unrelated to the local production, frequently visited the JV.[4]

Technology

The purpose of the JV formation was to make certain specialized communications equipment, namely, portable magnetor telephone switch boards for the Ministry of Defense. This product had been made for the last 15 years. During 1971 to 1973 Ericsson offered extensive training to the local technicians, both in the host country and in Sweden. According to the plan, technological responsibility was handed over to the local staff already at the end of 1974. The other activity which the JV had been carrying out increasingly for the last four years was to promote the transfer of know-how from Ericsson to other local companies. For this activity, the JV obtained its due commission.

In connection with the local production, the JV was in need of Ericsson's help to select new products. Whenever the JV learned of a demand for a certain related product, Ericsson was asked to supply all information regarding its manufacture so that a feasibility study could be made locally. Whenever necessary, Ericsson trained the local engineers in Sweden. The last training for such purpose was given to the local engineers in 1981. The JV enjoyed full freedom to undertake any action for its improvement.

Market

In marketing, the JV had two functions. One function was related to receiving orders from the buyers and delivering the goods after manufacture. The other was to act as liaison office for Ericsson to market the latter's products in the host country through the participation in international bids. Ericsson was heavily dependent on the JV for its second function.

Ericsson's contribution was important at the beginning insofar as it was required to show how the sales would occur. The local partner's role in selling the local products was very limited due to his ignorance thereof. The JV had no private buyer and produced mainly for the Ministry of Defense. There has been a substantial growth in sales in recent years.

The JV acted as local agent for Ericsson when tenders were submitted. The role of the JV was crucial because it maintained both formal and informal contacts with the buyers, supplied complementary information, and finally, attended the opening of the tenders. Notwithstanding the major role of the JV in trading, both Ericsson and the local partner had to make useful contributions in this respect. Ericsson's work consisted of preparing offers, arranging for the requisite products, passing information to the buyers through the JV and also visiting the buyers in case of need. The local partner often visited main customers and also provided legitimacy by his participation in the JV.

Resource Contribution Over Time

The contributions by the local partner, foreign partner and JV itself are summarized in Tables 3.3, 3.4 and 3.5 respectively. Immediately after the establishment of the JV, Ericsson assumed sole responsibility for deciding what materials were to be used, how many products were required and what sources would be utilized to

TABLE 3.3. Contribution of the Local Partner

Resource	Actual Contribution	Criticalness
Input Material	-	-
Capital	Cash investment on three occasions. Providing legitimacy to secure a loan.	Very Critical for the operating activity.
Manpower	Major activity during 1971 to 1973. Involvement in the appointment of key people.	Not critical.
Technology	-	-
Market	Visiting major customers.	Critical as legitimacy important.

procure them. However, already in 1974, a handful of local suppliers were enlisted and the JV could establish direct contact with the major external sources. The local partner made no contribution in this field.

Ericsson temporarily had a majority ownership in the JV al-

TABLE 3.4. Contribution of the Foreign Partner (Ericsson)

Resource	Actual Contribution	Criticalness
Input Material	Major contribution during 1971-1973.	Critical at the beginning.
Capital	No cash investment over the years.	Not critical.
Manpower	Foreign personnel's role during 1971-1973 and 1980-1985.	Critical for training and other activities.
Technology	Major contribution during 1971 to 1973. Technology transfer to local firms in 80's.	Critical for start-up and technical assistance.
Market	Supply of technical information to the buyers.	Critical for trading activity.

TABLE 3.5. Contribution of the JV Itself

Resource	Actual Contribution	Criticalness
Input Material	Major function from 1974.	Critical to manage own requirements.
Capital	-	-
Manpower	Main responsibility to inquire, select and appoint most of the employees after 1973.	Not critical.
Technology	Wholly responsible for local production between 1974 and 1985.	Critical to manage local production.
Market	Sale of the local products and a major role in trading business.	Critical as legitimacy and information about buyers vital.

though it did not invest any cash during the last 15 years. When the local partner was a minority owner (between 1971-1977), he made two useful contributions in this resource area; firstly, he provided cash to build up his share capital and secondly, he arranged local financing when the JV's own capital was not sufficient to undertake development programs.

In respect to manpower, Ericsson made some useful contributions at the beginning as it assigned local employees from its wholly-owned subsidiary and also sent its personnel to the host country to transfer technology and provide training to the local technicians. During this time and also afterwards, the local partner had the main responsibility for local recruitment. However, in 1976 the major task of recruitment came into the hands of the JV. In recent years the role of the foreign personnel have become important in connection with the sales of complicated projects to the local buyers.

Ericsson's role was vital between 1971 to 1973 in respect of manufacturing and training in the JV. From 1974 on, the local people managed the production activity and could also start R & D work later. Ericsson also performed technology transfer to other local firms (all government enterprises) during the 80s since there was restriction on private manufacturing of many items.

The JV was greatly assisted by Ericsson's old employees in marketing products in India, especially at the beginning. The JV was able to handle the whole marketing activities of its products alone from 1974 on. The JV also played a key role in trading, but Ericsson's active support was usually necessary to make the final deal. The local partner often had to visit prospective customers along with the GM and the foreign personnel to promote the products.

Control

Gradual change of share structure in favor of the local partner did not conflict with the interests of the foreign partner.[5] The formal control is shown in Table 3.6. There were four directors on the board. Two were nominated by Ericsson, while the other two were local shareholders. The local partner was chairman of the JV and the post of the GM was always left to a local business executive in

TABLE 3.6. Formal Control

Determinants of Control	Local Partner	Ericsson	JV	Criticalness
Shares 1971-75 1975-78 1978-86	(a) 26% (b) 40% (c) 60%	(a) 74% (b) 60% (c) 40%	-	Not critical.
Board	2	2	-	Not critical.
Chairman	The local partner throughout.	-	-	Critical to exercise formal control.
MD	-	-	A local business executive.	Critical for the development of the JV.
Plant Manager	-	-	Local experienced people.	Critical for the development of the JV.

order to emphasize efficiency and effectiveness in the company. This attitude of the owners enabled the JV to have formal control in many important areas. Competence was the major criterion in appointing plant managers.

The control through resources is illustrated in Table 3.7. As a buyer and a famous manufacturer of telecommunication equipment, Ericsson had enjoyed legitimacy for a great many years and could exercise some control through the procurement of input material. However, this control was short-lived, since the JV could meet all of its requirements from both local and external sources after a certain period.

The local partner had both formal and informal control over the financial activities of the JV. He worked as managing partner and had day-to-day contact with the venture. The GM often was required to consult with the local partner on financial matters and to obtain suggestions on how to spend the money. With a very few exceptions, all financial decisions were taken by the local partner during the last 15 years and his decisions were never challenged by the foreign partner.

The local partner could exercise control through the appointment of higher officials in the JV. However, the GM and the plant man-

TABLE 3.7. Control Through Resources

Resources	Local Partner	Ericsson	JV	Criticalness
Input Material	-	Control until the end of 1973.	Absolute responsibility during 1974 to 1985.	Not critical as a source of control.
Capital	Financial responsibility.	-	-	Very critical for the local partner's control.
Manpower	Control in major appointments.	-	Responsible for lower grade appointments.	Not critical in exercising control.
Technology	-	Responsibility until the end of 1973.	Operations during 1974 to 1985.	Critical for the JV's own control.
Market	Meeting with buyers.	Supply technical support to the large buyers.	Marketing of own products and supervising trading activity.	Some critical function by all parties but main control by the JV.

agers were also responsible for many appointments. The GM was free to choose efficient people, since the local partner always emphasized professionalism. Ericsson had no intention of retaining control in the technical area because the technology used in the JV was comparatively old. The JV succeeded in managing production, developing some new products and establishing a R & D department in the latter part of the operation. The JV could itself decide which products to manufacture, when to manufacture them and how the technology would be obtained.

From the beginning the JV had complete control over the marketing of its own products. It made contacts with the buyers, obtained orders, and finally supplied them on its own initiative. However, for the trading business, the JV had shared control with Ericsson, since it had to depend largely on Ericsson for offers, suggestions and technological information. Even the local partner enjoyed some control in marketing Ericsson's products, since he offered useful legitimacy and access to the many government buyers, which was essential in this context.

Cooperation/Conflict

In the JV the partners were very cooperative over the years. Ericsson always extended a helping hand to the local people, so the transfer of technology was possible within a very short time. Whenever it was needed, both the partners were willing to compromise. This attitude was always present and no substantial conflict could arise in the JV due to the candor of the partners. The local partner followed a professional kind of management and was aware of the foreign culture. Ericsson, the other partner of the present JV, had business activities worldwide through either wholly-owned subsidiaries or JVs. Ericsson's main representative and the local partner had been personal friends since the end of the sixties and they could easily solve most of the problems.

Some implicit conflicts were identified in the JV, however. There was a basic difference in opinion between the partners as to what would be its major area of interest. Ericsson was mainly interested in the trading business of the JV. The local partner's goal was always to produce more items locally and to gain recognition as a leading manufacturer of telecommunications components in the host country.

The local partner was somewhat suspicious of Ericsson's sincerity concerning the procurement of information. The JV needed information when selecting new products because it sought the rapid expansion of business through the increase of local production. It generally happened that such information arrived late or sometimes not at all. Another area of implicit conflict arose from the slow success of the JV in securing big local contracts. Ericsson openly blamed neither the JV nor the local partner for the failure but felt that someone was at fault.

On several occasions, the JV needed some concessions from Ericsson because it faced real problems and expected favors from the foreign partner. The local management tried to explain the situation and Ericsson could understand the problem. Concerning the conflict about local production, Ericsson was of the same opinion as the local partner. The foreign partner's representative asserted, "We understand that we need a big efficient local organization in order to win big contracts and to achieve that, we will probably have to go into more local production." The problem of responding to the JV's enquiries had been discussed several times in Ericsson.

Performance

From 1980 to 1985 the turnover of the JV was between 10 and 15M SEK, while the profit after taxes was 15% thereof. The main representative of Ericsson found the local management to be efficient and the collaboration between the partners excellent, but could not understand why the JV could not do more trading business. The local partner and the GM asserted that local products were doing very well in the Indian market. During the formation of the JV, the only manufacturing factory was established in Calcutta and the head office with the function of trading business was operating in New Delhi. At the end of 1980 a new factory was set up at Sahidabad Industrial Area, Gazibabad, U.P. where production activities and R & D works were pursued.

Concerning adaptation, the JV was successful in modifying many of the products to suit the Indian environment. Besides manufacturing and initiating direct sales from Sweden, the JV had also arranged transfer of know-how from Ericsson to the government firms in recent years, which the GM regarded as a case of adaptation. The participation of both the partners was important for the JV but neither needed to interfere in the day-to-day activities. Cooperation by both the partners and the JV was essential in connection with big trading business.

The JV has survived 15 years, in which no major crisis has arisen, which can threaten its existence.[6] The JV was exposed to some external threats, since it could not find more products for local manufacturing because of legal restrictions on many attractive items. There was some indication that the local authority could lift restrictions on some banned items for private manufacture. The GM expected that such a change would be very beneficial for the development and survival of the JV.

The local partner's interest in the JV increased over time. He raised his share of capital to become the main owner and also devoted more time and energy to the success of the company. To

Ericsson, the income from the JV had always been negligible compared with its other investments. But its interest in the JV was growing as it succeeded in signing some big technological contracts with the local government. Both Ericsson and the local partner saw the JV in a long-term perspective.

Conclusion

At the outset, Ericsson had formal responsibility to procure input materials and transfer technology to the JV. Ericsson succeeded to exercise control through these resources because it was balanced with the formal control. After three years, criticalness of both the resources decreased due to the availability of local material and ability of the local people in learning the production technique.

Ericsson never interfered in the day-to-day activities, since it relied wholly on the local management. The role of the local partner was that of supervisor and he had absolute control of the finance of the company. Both partners were cooperating and providing all possible support, so the JV relationship persisted to grow over time. This favorable situation prevailed for perhaps three reasons; firstly, both the partners were highly experienced, especially Ericsson; secondly, both partners had great interest in the JV, and finally, the main persons involved in the relationship were good friends. The JV enjoyed freedom to procure materials, to manage operations and to market the products. The controlling capacity in these vital areas allowed the JV to grow independently and to devote constant efforts to become further established in the host country.

MANAGEMENT IMPLICATIONS

We are aware that it is rather difficult to generalize the results and provide with some implications with the help of only one case study. However due to uniqueness of this study with its longitudinal approach, it has been possible to draw general as well as specific recommendations for partners.

General Implications

A major part of this study involves discussion and comparison between expected and actual contribution. When partners fail to assess each other's competence, there is a risk that primary estimation will be wrong and there will be a difference between expectation and actual contribution. The greater the difference, the higher the degree of disappointment of the partners. It is, therefore, essential to examine each other's capability carefully before the JV contract is signed.

This study suggests that an informal relationship between main partners solves problems in a JV. It is natural that many problems will arise but how partners will face them is always a vital question in JV. The first important step is that partners will not hide problems from each other, they will discuss the issues openly and with good intentions. These kinds of measures have been very helpful in avoiding conflicts in this study.

The reason to form a JV is to benefit from the support of the partners but it does not mean that the owners make unnecessary interference in the day-to-day activity. This study shows that it is wise to leave the management responsibility to the JV. Freedom of work by a JV minimizes suspicion between partners and leads to self-reliance. One finding of the study has been the differentiation between formal control and control through resources. Conflict can be caused when a partner has formal control but lacks a corresponding resource, or equally, when a partner provides a resource, even critical, but without formal control. The possibility of cooperation is positively related to the conformity between formal control and control through resources.

Implications for Foreign Partner

In a manufacturing JV, a foreign partner's main role is usually associated with the supply of technology and related information. The reason is that the local partner lacks technological knowledge and therefore depends on the other partner. The local partner is not usually able to assess the technical competence and the sincerity of the associate and this can create suspicion. This study suggests that in developing mutual understanding, a foreign partner should try to expedite the transfer of technology in local hands if the technology concerned is not so complicated and modern, and also supply of technical information whenever necessary. It is wise not to appoint foreign personnel just to exercise control in a JV. A foreign partner should see that expatriates are doing some important job which the local people cannot perform, otherwise there will be conflict sooner or later.

Learning is an important concept in a JV as the relationship develops over time. What the partners do not know in the early stages of a JV, they are able to learn due to the continuous exchange of resources. This study implies that the foreign partner's initial control changes and he is no longer in a position to exercise similar control over the years. If a local partner is competent in one field, he should be given responsibility for that job. It is more common that the local partner knows better about the local market, so the indigenous partner should have the responsibility for local marketing. This study suggests that involvement of local people assures legitimacy with the local buyers.

Implications for Local Partner

In most host countries, there is restriction on foreign investment, which means that a foreign firm cannot invest more than stipulated, even if the need is a critical one. A local partner needs to be financially sound and be able to keep his financial commitments. The local partner should try to create a good image for the financial institutions so that he can arrange money in case of need. When a local partner is responsible for local management, he should see that the foreign partner is well informed about the overall situation to avoid confusion. It is very easy for a foreign partner to become suspicious if there is a communication gap and the local partner has day-to-day control. This study suggests the above recommendations, inasmuch as the local partner managed to meet the requirements properly.

A common problem in a JV can be avoided if a local partner avoids nepotism in recruiting local employees. This study suggests that the local partner's professional attitude created reliance with the foreign partner. A local partner can try to have a long-term perspective to match with the goal of the other partner. This is also positive for developing a relationship over time. There is a necessity that a local partner is on good terms with the local government officials, but too much involvement with them may create suspicion. A local partner should openly discuss his contacts with the authorities with the foreign partner and also the purpose for those contacts, if the topic is related to the JV's interests. In this JV, the local partner tried to understand the needs of the other partner and also kept the latter informed about what was going on in the JV.

REFERENCE NOTES

1. The data were actually collected in connection with Syed Akmal Hyder's thesis, "The Development of International Joint Venture Relationships: A Longitudinal Study of Exchange of Resources, Control and Conflicts" (1988).
2. See March and Simon (1958), Walton and McKersie (1956) and Kochan, Huber and Cummings (1975).
3. Input materials were mainly different kinds of components to produce specialized communication equipment.
4. This kind of visit by the foreign personnel increased recently in connection with technology transfer to the local government firms as part of big contracts. These personnel came from different units of Ericsson in contrast to the visits during the formation of the JV.
5. One executive of Ericsson commented:

> Our aim is to build up an organization in a right way so that it can represent Ericsson satisfactorily. There are many companies around the world who need either majority ownership or no ownership at all. But our case is different. We have government as our main customer, so we must be flexible concerning ownership and comply with the desires of the host country.

6. Fifteen years have been calculated from 1971 to 1985. The JV was also operating at the end of 1987 but the present discussion covers the period up to 1985.

REFERENCES

Abdul, A.R. (1979). *The Mixed Enterprises in Malaysia: A Study of Joint Ventures Between Malaysian Public Corporations and Foreign Enterprises*, Dissertation No. 36, Katholieke Universiteit te Leuven.

Ahn, Doo-Soon (1980) "Joint Ventures in the ASEAN Countries," *Intereconomics*, July/August, pp. 193-198.

Aiken, M. & Hage, J. (1968). "Organizational Interdependence and Interorganizational Structure," *American Sociological Review*, 63, pp. 912-930.

Barlew, F.K. (1984). "The Joint Venture – A Way into Foreign Markets," *Harvard Business Review*, 62, July-August, pp. 48-54.

Beamish, P.W. (1985). "The Characteristics of Joint Ventures in Developed and Developing Countries." *Columbia Journal of World Business*, 20, Fall, pp. 13-19.

Beamish, P. & Lane, H.W. (1982). "Joint Venture Performance in Developing Countries," ASAC conference, Ottawa University.

Berg, S.V. & Friedman, P. (1980). "Corporate Courtship and Successful Joint Ventures," *California Management Review*, 22, No. 2, pp. 85-91.

Bivens, K.K., & Lovell, E.B. (1966). *Joint Ventures with Foreign Partners*. NY: National Industrial Conference Board, Inc.

Campbell, J.P. (1977). "On the Nature of Organizational Effectiveness," In *New Perspectives of Organizational Effectiveness*, Goodman, P.S. & Pennings, J.M. (eds.), San Francisco: Jossey-Bass, pp. 13-55.

Connolly, S.G. (1984). "Joint Ventures with Third World Multinationals: A New Form of Entry to International Markets," *Columbia Journal of World Business*, 19, Summer, pp. 18-22.

Dunning, J. (1978). "Eclectic Theory of International Production, Organizational Forms and Ownership Patterns." Working paper, University of Reading.

Edström, A. (1975). "The Stability of Joint Ventures." Re-Rapport, Företagsekonomiska Institutionen, Göteborg: Göteborgs Universitet.

Franko, L.G. (1971). *Joint Venture Survival in Multinational Corporations*. NY: Praeger Publishers.

Friedmann, W. & Béguin, J-P. (1971). *Joint International Business Ventures in Developing Countries*. NY: Columbia University Press.

Ghauri, P.N., *Negotiating International Package Deals*. Acta Universitatis Upsaliensis, Studia Oeconomiae Negotiorum, No. 17, (distributed by Almqvist & Wiksell International, Stockholm).

Gullander, S. (1976). "Joint Ventures and Corporate Strategy." *Columbia Journal of World Business*, 11, Spring, pp. 104-114.

Håkansson, H. (eds.) (1986). *Industrial Technological Development: A Network Approach*. Kent: Croom Helm Ltd.

Harrigan, K.R. (1984). "Joint Ventures and Global Strategies." *Columbia Journal of World Business*, 19, Summer, pp. 7-16.

Harrigan, K.R. (1986). *Managing for Joint Venture Success*. Lexington: Lexington Books, D.C. Heath and Company.

Janger, A.R. (1980). "Organization of International Joint Ventures." Conference Board Report No. 787, NY.

Killing, J.P. (1982). "How to Make a Global Joint Venture Work." *Harvard Business Review*, 60, May-June, pp. 120-127.

Killing, J.P. (1983). *Strategies for Joint Venture Success*. Kent: Croom Helm Ltd.

Koot, W.T.M. (1986). "Underlying Dilemmas in the Management of International Joint Ventures." Paper presented at the Conference of Cooperation Strategies in International Business, The Wharton School and Rutgers Graduate School of Management, October.

Larsson, A. (1985). *Structure and Change. Power in the Transnational Enterprise*. Acta Universitatis Upsaliensis, Studia Oeconomiae Negotiorum, No. 23, (distributed by Almqvist & Wiksell International, Stockholm).

Otterbeck, L. (1979). "Joint Ventures on Foreign Markets." Stockholm School of Economics (mimeo).

Otterbeck, L. (1980). "Management of Joint Ventures." RP 80/10, Institute of International Business, Stockholm School of Economics.

Pfeffer, J. (1978). *Organizational Design*. IL: AHM Publishing Company.

Pfeffer, J. (1981). *Power in Organizations*. MA: Pitman, Marshfield.

Pfeffer, J. & Nowak, P. (1976). "Joint Ventures and Interorganizational Interdependence." *Administrative Science Quarterly*, 21, No. 3, pp. 398-418.

Pfeffer, J. & Salancik, G.R. (1978). *The External Control of Organizations*. NY: Harper and Row.

Raveed, S.R., & Renforth, W. (1983). "State Enterprise – Multinational Corporation Joint Ventures: How Well Do They Meet Partners' Needs?" *Management International Review*, 23, No. 1, pp. 47-57.

Reynolds, J.I. (1984). "The 'Pinched Shoe' Effect of International Joint Ventures." *Columbia Journal of World Business*, 19, No. 2, pp. 23-29.

Schmidt, S.M. & Kochan, T.A. (1972). "Conflict: Toward Conceptual Clarity." *Administrative Science Quarterly*, 17, No. 3, pp. 359-370.

Simiar, F. (1983). "Major Causes of Joint Venture Failures in the Middle East: The Case of Iran." *Management International Review*, 23, No. 1, pp. 58-68.

Steers, P.M. (1977). *Organizational Effectiveness: A Behavioral View*. Santa Monica, CA: Goodyear.

Tomlinson, J.W.C. (1970). *The Joint Venture Process in International Business: India and Pakistan*, Cambridge, MA: M.I.T. Press.

Vernon, R. (1972). *Restrictive Business Practices: The Operations of Multinational United States Enterprises in Developing Countries: Their Role in Trade and Development*. NY: U.N.

Walmsley, J. (1982). *Handbook of International Joint Venture*. London: Graham & Trotman, London.

Chapter 4

New Market Entry: Exporter Experience and Government Support

F. H. Rolf Seringhaus

INTRODUCTION

Export behavior and activity has received a great deal of attention in the literature. Frequently such research has dealt with the difficulties and complexities firms encounter without paying much attention to the particular stage of their foreign market involvement. The model of exporting process is now widely accepted among export researchers, as is the view that one of the most critical phases in this process is the market entry. Management know-how and experience to compete internationally is generally regarded as a vital part in the exporting process (Weekly and Bardi, 1975/76, Bilkey and Tesar, 1977). For the new exporter, acquisition of such knowledge may be quite difficult and help is often sought outside of the firm. For the experienced exporter, previous market entries serve as a significant source of knowledge, however, new markets also pose new problems and demand new solutions and approaches. The situation is one where, on the one hand, the firm needs to build up required skills, and on the other hand, it needs to use its adequate existing skills to overcome the challenges posed by a new market entry.

One important source of support and know-how for exporting lies in various programs governments have developed to promote exports generally and provide specific help to firms becoming involved in foreign markets. Several recent articles focus on the role and impact of such government support (Seringhaus, 1986, 1987;

Seringhaus and Mayer 1988; Denis and Depelteau, 1985). Research generally has not distinguished between new and experienced exporters; Czinkota's (1982) examination of difficulties encountered by exporters at different export stages is a notable exception. The purpose of this article is to analyze the decisions and problems experienced exporters face in the entry of new markets and the role governmental export support plays in the market entry process. The approach taken in this study is based on an empirical model rather than on a normative model of export behavior (such as those suggested by Root, 1966, Grub, 1971, or Dichtl et al., 1984). It is thus suggested that an export behavior framework should be based upon positive firm behavior.

This article will first provide a conceptual basis and a research and sample methodology, followed by characteristics of new market entry decisions, problems faced by experienced exporters, and what their usage of government support is. Then decision aspects and entry problems are related, and, finally, implications are derived from the market entry experience and the role of export support, and possible steps exporters can take are discussed.

CONCEPTUAL BACKGROUND

Often, a firm's decision to export and the decision to enter new export markets are not distinguished in investigations of export management behavior. When a firm decides to become an exporter, its goals and objectives, indeed every operational aspect of the firm, will be affected by this reorientation. On the other hand, when established exporters enter new markets, the "learning by doing" dictum is relevant. This means that the decision to enter a new market may appear easier since it has been done before, and one expects that the firm should have learned from previous problems. Equally relevant, however, is the view that new market entry by experienced exporters is not routine or problem-free. First-time exporters lack such experience and have no yardstick for the decisions which bring about an operational export commitment for the firm. In general, the literature is lacking in focused examination of the new market entry experience of established exporters.

While, in this paper, the emphasis is on the market entry experience of the firm, the concept of government export support needs a brief comment. Many governments, at the federal, state or provincial level, offer export support programs to their respective business communities. This support often ranges from information, statistics and advice on markets and opportunities to programs whereby firms receive organizational and financial support to visit foreign markets and participate in international trade fairs or exhibitions. Often research on the role of such support in the firm has been limited to measuring awareness and usage levels of programs, but recently, studies concerned with the impact of government support on firm performance have become available (Seringhaus, 1986). Thus, experienced exporters, by definition, possess a substantial body of expertise that includes knowledge of various sources of know-how such as government export support.

An analysis of new market entry decisions and problems should reveal which decision elements are salient and where firms might have to change their behavior, that is, adjust to the idiosyncracies of each individual export market. Johanson and Vahlne (1977) have suggested that knowledge from earlier experience is indeed situation-specific. The implication is that what has been learned in one market may not be suitable or sufficient for another market. Problems observed for experienced exporters then may point to the need for continuous learning and rebuilding of marketing knowledge. Weaknesses in export marketing suggest that, perhaps, a new approach to assessing a firm's export competence is needed. In addition, given the availability and use of governmental export support, questions may be raised about the adequacy and content of such programs.

RESEARCH AND SAMPLE METHODOLOGY

Proximity and cultural similarity to the U.S.A. are often powerful inducements for Canadian firms planning to export. The perception that export involvement in the U.S. prepares a firm for later export involvement in overseas markets is only partially true, because no other export market is as similar to the Canadian market as the U.S. market is. By contrast, most overseas export markets have substantially different environments, business practices, political and regulatory settings.

Export market entry is acknowledged to present a greater challenge for small- and medium-sized firms, who are more resource-constrained and find it more difficult to obtain exporting know-how, than for larger firms. The target group chosen for this study is small- and medium-sized experienced exporters who recently entered a new overseas export market.

A systematic random sample with firm characteristics of 50 to 500 employees, single division organizational structure from the export-intensive industry sectors of machinery, electric/electronic products and fabricated metal, consisting of 128 Canadian firms was selected. Data were collected from 90 participants (70.3% response rate) through personal interviews using a structured questionnaire. Table 4.1 provides profile data of the sample.

NEW EXPORT MARKET ENTRY DECISIONS AND PROBLEMS

As a backdrop to the discussion of exporters' decisions and problems it is interesting to note that about two-thirds (68%) of firms carry on their exporting activities as an integral part of general marketing efforts, while the remainder operate with a separate export department. These experienced exporters have fairly clear export policies. Two-thirds have a declared policy of unlimited export expansion, in other words, growth through exports wherever opportunities exist. Firms require the necessary confidence and competence to pursue this type of policy. One-third, however, indicated that their approach was more controlled, that is their expansion is limited by specific objectives and constraints.

TABLE 4.1. Profile of Respondent Firms

- Average firm size 140 employees
- Average total sales $12.9 million
- 90% have been established longer than 10 years
- 2/3 have been exporting between 2 to 5 years
- Exporting to an average of 7 overseas markets
- Average share of total sales exported 44.7%
- Average export experience of respondent 10 years

Most firms also engage in management planning for export (92%). Some 81% systematically search for new markets. This means screening markets to single out those with export potential. Seventy-three percent carry out research on such potential markets to enable the exporter to make a selection decision upon which later market entry may be based.

The process of foreign market involvement is complex and usually requires management to make a number of decisions as well as carry out activities which, in retrospect, appear like distinct decisions when they actually were a stream of more or less sensible actions. It is difficult to separate managing and decision-making since they are often interwoven (Simon, 1960). To some extent the latter explains why business decision-making often cannot be observed and must be inferred from management activities. One approach to assist in this dilemma is to break the export management process down into smaller components. In this study, respondents were asked about the importance of a variety of decision aspects as they relate to new market entry. Table 4.2 gives a ratio of the proportion of firms viewing a decision aspect important versus those viewing it as unimportant.

The proportion of firms mentioning export profit expectation and management interest in entering the new market as having an important bearing on the decision outnumbered those that did not 17 to 1. By contrast, most firms saw protected technology (such as product patent) as unimportant. In a similar vein, problems encountered in this market entry are also shown in Table 4.2. The ratio is the proportion of firms experiencing a problem versus those who did not. The most frequently noted problem was that of competition in the market while product standards and certification requirements were the least problematic. New market entry problems are specific and actionable, that is, an exporter might respond to a noted problem with concrete action. For instance, the high cost of selling to an export market may prompt a review of current practice, resulting in such changes as: local agents are used in customer calls rather than head-office personnel, shipping arrangements may consider intermodel transport, promotional activities may be carried out in a more cost-efficient manner using greater standardization, and so on. We

TABLE 4.2. Decisions and Problems of New Market Entry

Aspects of Entry Decision[1]	Ratio of Proportion Important to Unimportant
Expected Profits	17.2
Management Interest	16.8
Active Market Investigation	3.1
Product Price	3.1
Limited Canadian Market	2.7
Marketing Expertise	2.1
Financial Resources	1.4
Promotion in Export Market	.8
Unsolicited Order	.6
Protected Technology	.2

Aspects Creating Problems[1]	Ratio of Proportion Problem to No Problem
Competition in Market	2.9
Expense of Dealing in Market	1.5
Distance of Market	1.5
Representation in Market	1.0
Reliability of Market Data	.9
Government Policy in Market	.8
Language	.7
Business Practices	.7
Currency Exchange Regulations	.4
Product Standards, Certification	.3
	(n = 90)

[1] Measurement is on 3-point scale; for decision aspects: very important, somewhat important, not important; for problem aspects: definite problem, some problem, no problem; the respective first two categories are combined and expressed as ratio of third category.

now turn to the role governmental export support plays in these firms by reviewing the extent of awareness and usage.

As mentioned earlier, experienced exporters have a broad base of expertise. Table 4.3 provides evidence that few experienced exporters are unaware of the export support available from government. Indeed, about one-half or more of the respondent firms have used governmental export support. This support falls into two broad categories. One type is informational in nature, such as statistics, details on markets, advice given through government offices (foreign as well as domestic). The other type is hands-on and contact-facilitating as in cost-sharing of market visits, joining of organized trade missions to foreign markets, or financial and/or organizational support for participating in international trade fairs. Nearly two-thirds of firms used government programs to visit export markets. Somewhat more than one-half joined others in group trade missions, while somewhat fewer than one-half made use of trade fair support and the information services provided through various offices.

Having seen the decisions and problems experienced by the exporters and the fairly extensive usage of government support, the obvious question is: what difference, if any, did this help make? Table 4.4 offers some insight into the impact of support on different export marketing activities. Measurement of the impact of support on firm activity is complex and difficult (Seringhaus and Mayer, 1988) and Table 4.4 gives the extent to which government support used by a firm contributed to such activity. Two observations stand out. First, support was most helpful in establishing market contact through visits and actual market entry. This seems determined by the high proportion of firms using contact services. Second, knowledge of markets and opportunities appears to have benefited, while the actual task of planning and readying for export have not. The latter suggests that the services of government offices may not be the kind firms actually need.

TABLE 4.3. Awareness and Use of Government Export Support

Type of Export Support	Used Support	Aware of Support	Unaware of Support
Market visits	64.4%	25.6%	10.0%
Trade missions to markets	54.2	38.5	7.3
Trade fair participation	45.7	45.0	9.3
Government offices	47.5	39.7	12.8

(n = 90)

Note: Responses cover identical or similar support services provided by different government programs or by different levels of government.

Next is an attempt to interpret the decision and problem aspects in light of the contribution of government support to the exporting activity associated with market entry.

Management concern with the investigation of markets and marketing expertise suggests that it is very much aware of the need for competence in that area. Extensive use of government support furthermore suggests that some incremental benefit could be expected therefrom. The problems experienced reveal a less comforting picture. While judgement that the firm needed support was right, the benefits—beyond visiting and entering the market—to the marketing effort were questionable. The latter is borne out by the barriers (viz. competition, expense, market access) firms experienced in the market despite government support. For this group of exporters, one may conclude that some of the support services might be unsuitable to the experienced exporter, or not sufficiently tailored to the specific needs of these firms. Furthermore, the fact that exporters are experienced does not obviate the need for, and usefulness of, appropriate government export support.

We now return to the decision and problem factors in order to further develop our understanding of the firms' market entry experience.

As a next step the decision and problem factors were factor analyzed* and each reduced to four factors. Table 4.5 shows the factors

TABLE 4.4. Contribution Made by Government Export Support

Export Marketing Activity	Substantial Contribution	Some Contribution	No Contribution
Plan export marketing	11.6%	32.6%	55.8%
Assess risks and uncertainties	14.0	44.2	41.8
Identify opportunities	29.1	44.2	26.7
Knowledge of markets	36.1	45.3	18.6
Prepare for exporting	12.8	43.0	44.2
Visit markets	53.5	31.4	15.1
Financial feasibility of exporting	17.4	29.1	53.5
Enter new markets	40.7	50.0	9.3

(n = 90)

*Factor Analysis, Principal Component Method, Equamax Rotation

TABLE 4.5. Factor Analysis of Decision and Problem Variables in New Market Entry

	Variables Represented	Factor Loading	Eigen Value	Variance Explained
Decision Factors:				
EFFORT	Marketing Expertise	.80	2.88	28.8%
	Market Investigation	.80		
	Financial Resources	.67		
	Promotion in Market	.61		
PASSIVE	Protected Technology	.77	1.62	16.2%
	Unsolicited Order	.75		(44.9%)
DRIVE	Limited Home Market	.86	1.17	11.7%
	Management Interest	.54		(56.6%)
FINANCIAL	Product Price	.87	1.00	10.0%
	Expected Profits	.56		(66.6%)
Problem Factors:				
EXPERTISE	Language	.76	2.42	24.2%
	Representation	.66		
	Government Policy	.66		
	Business practices	.45		
REACH	Distance	.80	1.69	16.9%
	Expense	.77		(41.2%)
MARKET	Competition	.79	1.30	13.0%
	Market Data	.74		(54.2%)
BARRIER	Exchange Regulation	.77	1.00	9.8%
	Product Standards	.69		(64.1%)

and the variables they represent. Effort emerged as the most important factor in the entry decision. A firm's export marketing effort and commitment, as in the case of new market entry, is also a function of its prior exporting experience. Effort by experienced firms differs from that of new exporters discussed in the literature. The latter face choices of direct or indirect exporting or major or minor commitment of resources (Bilkey and Tesar, 1977, Brasch, 1981). Also, feedback during early export involvement is very important to the new exporter. If such feedback is negative, it may adversely affect the firm's effort and ultimately result in export failure (Welch and Wiedersheim-Paul, 1980). Both experienced and new exporters, however, gear their effort and commitment to their perception of export market opportunities.

The factor EXPERTISE is the most significant of the four problem factors and it reveals that experienced exporters are not exempt from learning and adaptation to the challenges of new markets. Several researchers have reported on exporting problems similar to those listed in Table 4.2, however, generally not in a specific market entry context. What the foregoing suggests is that export knowledge and expertise are not only linked to perceived exporting problems (Weekly and Bardi, 1975/76, Pavord and Bogart, 1975, Simpson and Kujawa, 1974, Roy and Simpson, 1981) but, perhaps more significantly, to actual problems experienced.

Before discussing the relationship between decision and problem factors, the possibility that firm responses could be influenced by the market region in which the entry occurred needs to be recognized. Manova analyses were carried out on the decision and problem factors for the five regions (South and Central America, Western Europe, Asia, Australia and New Zealand, Middle East and Eastern Europe). No significant variance across these regions was noted, thus permitting a generalized discussion of decision and problem aspects.

ANALYSIS OF THE DECISION-PROBLEM RELATIONSHIP

Earlier it was seen that exporters encountered a particular pattern of market entry problems, given a certain pattern of entry decision aspects. Based on the strength of the finding that approximately 40% of bivariate associations between decision and problem variables are statistically significant, Canonical Correlation was used to investigate the strength and form of the relationships.*

Canonical Correlation facilitates the exploration of interrelations among sets of multiple criterion variables and multiple predictor variables. The premise used here is that exporters' decision aspects may have some bearing on the actual market entry problems encountered. Hence the entry problems comprise the criterion group

---

*Only individual variables in one group (i.e., Decision aspects) showing statistically significant correlation across at least fifty percent of individual variables in the other group (i.e., Problem aspects).

and the decision aspects make up the predictor group. While the Canonical Correlation technique generates as many functions as there are criterion variables (six in this example) only the first three functions will be interpreted. Redundancy analysis (Table 4.6) of criterion and predictor sets shows that the first two functions account for most of the proportion of the total redundancy (i.e., .858 for the criterion set, and .879 for the predictor set) and thus are most important to understanding the relationship. In addition, the mean square of Canonical Correlation (MSCC) of .20, for the two functions is statistically significant.

Since interpretation of Canonical Correlation is complex, three measures of the contribution of each variable to the canonical relationships are provided. Table 4.7 includes standardized canonical weights which predict the relative importance of variables in the overall relationship. Loadings give the correlation between individual variables and the function (i.e., canonical variate). Finally,

TABLE 4.6. Redundancy Analysis

Canonical Function	R^2	VE	Redundancy	Proportion of Total Redundancy
Problems:				
Criterion Set 1	.26	.21	.0546	.572
2	.13	.21	.0273	.286
3	.08	.13	.0104	.109
			.0954	
Decisions:				
Predictor Set 1	.26	.31	.0806	.757
2	.13	.10	.0130	.122
3	.08	.13	.0104	.097
			.1064	

Note: VE = Variance extracted equals canonical loadings squared
Redundancy = $(R^2)(VE)$

TABLE 4.7. Relationships Between Variables and Canonical Functions

	Function 1			Function 2			Function 3		
Problem Variables	Weight	Loading	%L²	Weight	Loading	%L²	Weight	Loading	%L²
(Criterion Set)									
Gov't Policy	-.82	-.82	54.8	.02	-.23	4.1	-.49	-.30	12.1
Market Data	.29	-.13	1.4	-.94	-.88	60.0	.12	.29	11.3
Representation	-.03	-.36	10.6	-.09	-.39	11.8	.45	.39	20.4
Distance	.21	-.13	1.4	-.40	-.43	13.0	-.85	-.55	40.6
Competition	-.57	-.48	18.8	.41	-.08	.5	.32	.26	9.1
Expense	-.28	-.40	13.0	-.02	-.37	10.6	.44	.22	6.5
			100.0			100.0			100.0
Decision Variables									
(Predictor Set)									
Promotion in Market	.38	.01	0	.01	-.25	10.4	-.62	.43	23.6
Market Investigation	-.41	-.69	26.0	-.30	-.24	9.6	-.81	-.33	13.9
Management Interest	-.59	-.82	36.7	.18	-.04	.3	.74	.53	35.8
Marketing Expertise	-.37	-.64	22.3	1.07	.06	.6	-.10	-.02	.1
Financial Resource	.01	-.45	11.0	-1.29	-.64	67.9	-.06	.08	.8
Expected Profits	-.002	-.27	4.0	-.15	-.26	11.2	.18	.45	25.8
			100.0			100.0			100.0
Canonical R		.51			.36			.28	
Root		.26			.13			.08	

Note: %L² = Loading squared and expressed as a percentage of the sum of squared correlations (loadings) each variate.

loadings squared and expressed as a percentage of their sum provide another way of ranking the contribution of variables. These, in conjunction with the redundancy measure provide a sufficient basis for interpretation of the functions (Alpert and Peterson, 1972).

There are some differences in the rank order of the variables of functions. In the first function, the order in which market entry problems contribute to the relationship as suggested by the weights is Government Policy, Competition, Market Data, Expense and Distance. Using statistically significant loadings, the rank order is Government Policy, Competition, Expense and Representation. The percentage of variate communality is more sensitive to variations in contributions of variables and suggests that primarily Government Policy, followed by Competition and Expense are related to decision aspects Management Interest, Market Investigation and Marketing Expertise. Thus, it seems that market environment (regulatory) and market dynamics (competition and cost of marketing) stand out as problem aspects. Management's keenness while emphasizing market analysis and availability of in-house expertise may have rushed the process of the market entry or underestimated the complexity of the market. Thus, limited usefulness of prior experience and the impact of environment and dynamics encountered in the market on the exporter's marketing activities are underlying dimensions of the market entry.

The second function (using variate communality in the interpretation) shows problems of Market Data, Market Distance related to primarily one decision aspect, namely Financial Resources. Resources are universally critical to the firm and, in the context of exporting, a reflection of management's commitment. The relationship above suggests that the search for and acquisition of appropriate market data, perhaps, was constrained by limited resources, or less successful than had been expected. The reliability of market data thus has potential ramifications on a broad scale as far as an exporter's market entry is concerned.

The third function suggests that the distance of the export market and, to a lesser extent, representation problems are related to management interest and profit expectations relative to the market entry. Underlying this, albeit less important relationship, may be the realization that physical distance, control and costs should influence a firm's market choices.

The relationship between the problems experienced by exporters and the decision characteristics preceding market entry may be summarized as follows:

Problem Area	Decision Area
primarily related to market environment and dynamics	primarily related to commitment and attitude
• Government policy	• Management interest
• Competition	• Market investigation
• Expense of marketing	• Marketing expertise
• Market data reliability	• Financial resources
• Distance of market	• Profit expectation

At the risk of oversimplification, the market entry difficulties encountered suggest (1) that market entry was rushed, (2) the complexity of the market was underestimated, and (3) the data search and acquisition were insufficient.

CONCLUSIONS AND IMPLICATIONS

This study seeks to expand the export behavior literature by focusing on the decision and problem characteristics surrounding export expansion of Canadian firms through new overseas market entry. Much of the existing literature deals with export decisions and perceived problems in a general way. The contribution of this research lies in the analysis of actual market entry experience, that is, problems encountered and their relation to decision facets associated with entry. The usage of government support to exporting activity, interpreted in light of entry experience, permits some inferences as to the utility and appropriateness of such support.

Factors of importance in the decision to enter a new market show that effort, manifested by several specific decision facets, is critical to such export expansion. Recognizing particular decision facets as important, however, hardly suffices to make the entry successful. It is suggested that competent implementation of such decision facets is vital. In other words, export marketing practice must ensure that the market entry occurs as anticipated. This, in turn, depends on a firm's export marketing experience and expertise. Knowledge and use of government support services are seen as part of such know-how and the extensive use made of them suggests that firms expected to gain some incremental benefit.

Market entry experience showed that the most significant problems are skill-related and concern export marketing competence. Indeed, a number of the entry problems reported seem to originate, to some extent, in the implementation of export marketing activities. It is noted that government support makes its main contribution in actual contact with a market and thereby benefit most through increasing a firm's knowledge of a market. Planning and preparation is largely in the hands of the firm itself and government support seems of little consequence. Competent implementation of export marketing effort, as seen by entry problems, such as competition, also remains beyond the influence of government support.

These results lend empirical support to the contention that experiential knowledge may not be applicable across export markets and that each market entry not only requires thorough preparation but also the acquisition of new knowledge. In addition, experienced exporters find greatest benefit from government support in a narrow part of the exporting process. This leaves unanswered questions such as: Do firms prefer to handle preparation and post-contact marketing efforts on their own? Are their needs too specific for available government programs? Do existing programs not differentiate the needs of different types of firms? Are programs of limited scope and benefit for experienced exporters?

The observations, implications and required actions can be drawn together in the following way:

Observation:	Implication:	Management Action:
•Problems are skill-related	•Each market entry requires thorough preparation	•Review entry experience with learning focus
•Prior experience of limited use	•Implementation of decision is critical	•Plan and prepare for entry on realistic interpretation of indicators
•Quality of pre-entry and entry action affects overall entry experience	•Export marketing competence is dynamic	•Pragmatic and objective analysis of resource supply and demand
	•Prior experience helpful but not sufficient for export expansion	

It is clear that entering new foreign market environments is challenging even for experienced exporters and that a myopic view, flowing from previous exporting successes, may hamper a firm's competitive competence. Prior export market entry experience is a useful base upon which to build further export expansion, however, it is equally important to realize that such expertise may not be sufficient for new markets. It is also noteworthy that government support can make a useful contribution to at least a part of exporting activity. Perhaps it should be made clear by program managers that support is appropriate in a highly focused sense, that is, help with market contact, for example, and not as an umbrella service extending over most of the exporting process. A brief comment is in order on the three recommendations for management action. First, careful and constructive review of market entry experience should be undertaken with a focus on problem recognition and improvement. This involves assessing whether market research produced reliable data on the competitive market setting and the regulatory environment with its potentially inhibiting or supportive effect on the market. Second, planning of new market entries should be based on weighing carefully which past experience may be useful and what uncertainties exist where new knowledge is needed. Market indicators need to be interpreted realistically and contingency scenarios included in the entry preparations. Third, an objective evaluation of the firm's exporting competence and weaknesses in skills and expertise should be a periodic undertaking, and will go a long way toward reducing or eliminating potential problems in the market entry process. Government support should be carefully screened and selective use be made of those services that can tangibly contribute to greater export marketing effectiveness. Overall, it is export management's task to make new market entry less problem-prone, although it would be unrealistic to expect entry to be problem-free.

REFERENCES

Alpert, M.I. & Peterson, R.A. (1972). "On the Interpretation of Canonical Analysis," *Journal of Marketing Research*, 9, pp. 187-192.

Bilkey, Warren J. & Tesar, George (1977). "The Export Behavior of Smaller Sized Wisconsin Manufacturing Firms." *Journal of International Business Studies*, Vol. 8, No. 1, Spring/Summer, pp. 93-98.

Brasch, John J. (1981). "Deciding on Organizational Structure for Entry into Export Marketing." *Journal of Small Business Management*, Vol 19, No. 2, April, pp. 7-17.

Czinkota, Michael R. (1982). *Export Development Strategies.* NY: Praeger Publishers.

Denis, Jean-Emile & Depelteau, Daniel (1985). "Market Knowledge, Diversification and Export Expansion." *Journal of International Business Studies*, Vol. 16, Fall, pp. 77-89.

Dichtl, E., Leibold, M., Koeglmayr, H.G. & Mueller, S. (1984). "The Export Decision of Small and Medium-Sized Firms: A Review." *Management International Review*, Vol. 24, No. 2, pp. 49-60.

Johanson, Jan & Vahlne, Jan-Erik (1977). "The International Process of The Firm—A Model of Knowledge Development and Increasing Foreign Market Commitment." *Journal of International Business Studies*, Vol. 8, Spring/Summer, pp. 23-32.

Pavord, William C. & Bogart, Raymond G. (1975). "The Dynamics of the Decision to Export." *Akron Business and Economic Review*, Vol. 6, No. 1, Spring, pp. 6-11.

Root, Franklin R. (1966). *Strategic Planning for Export Marketing.* Scranton, International Textbook Company.

Roy, Delwin A. & Simpson, Claude L. (1981). "Export Attitudes of Business Executives in the Smaller Manufacturing Firms." *Journal of Small Business Management*, Vol. 19, No. 2, April, pp. 16-22.

Seringhaus, F.H. Rolf (1986). "The Impact of Government Export Marketing Assistance on Firms." *International Marketing Review*, Vol. 3, No. 3, pp. 55-56.

——— (1987). "Export Promotion: The Role and Impact of Government Services," *Irish Marketing Review*, Vol. II, Spring, pp. 106-116.

——— and Mayer, C.S. (1988). "Different Approaches to Foreign Market Entry Between Users and Non-Users of Trade Missions." *European Journal of Marketing*, Vol. 22, No. 10, pp. 7-18.

Simpson Jr., Claude L. & Kujawa, Duane (1974). "The Export Decision Process: An Empirical Enquiry." *Journal of International Business Studies*, Vol. 5, No. 1, Spring, pp. 107-117.

Simon, Herbert (1960). *The New Science of Management Decision*. NY: Harper and Row, Publishers.

Weekly, James K. & Bardi, Edward J. (1975/76). "Managerial Perceptions of Exporting Problem Areas." *Baylor Business Studies*, Vol. 6, No. 4, pp. 17-26.

Welch, Lawrence S. & Wiedersheim-Paul, Finn (1980). "Initial Exports—A Marketing Failure?." *The Journal of Management Studies*, Vol. 17, No. 3, October, pp. 332-344.

SECTION III: SOCIOPOLITICAL INTERNATIONAL MARKETING ISSUES

Chapter 5

Learning from Corporate Leaders: A Modern Sociocultural Development of International Management Interactions in the Philips, Holland

Nico J. Vink

NEEDED: A DYNAMIC PARADIGM

Introduction

This treatise deals with the question of decision making processes in situations of external and internal change of managements at different levels of multinational corporations, and the importance of teaching and learning on the way. The first section deals with theory, the second one with applying theory to the actual practices within the Philips corporation in the sixties to eighties.

Our basic assumption is that most theoretical paradigms are not adequately fit for analyzing and managing decision making processes in situations of major change. We need a new theoretical composite model. In order to develop such a model, we will look at some key paradigms in theories of organization, management and marketing. We will also look into social responsibility and business ethics thinking, because so many businesses have been blamed for not responding properly to external changes.

Organization Theory

With Simon and March (1957 and 1958) organizational decision making has been virtually reduced to individual managers freely choosing problems and solutions, however limited the number of alternatives, and the "bounded rational" ways in which these alternatives are chosen, may be. Conflicts may hamper rationality, but they are only considered to be exceptions to the rule of harmony.

In Cyert and March's thinking (1963) corporate behavior gets prime focus. Corporations are managed by voluntarily chosen shifting coalitions of actors. In this more realistic approach, each management level has its own environmental uncertainties and perceptions. Conflicting interests are normal and decision making is therefore very much a political process. Decision making is also considered to be a form of organizational learning. They also made it clear how companies respond to uncertainty. In spite of this, however, "power" and "morale" are still being treated in a "black box" way. Moreover, they greatly seem to overestimate corporate freedoms of choice and responsiveness. Actor relationships are treated like logics and rational divisions of labor, not as social processes.

According to system theoreticians like Emery and Trist (1965, pp. 21-32) corporations react to uncertain market conditions or outside "stakeholders" (1972). Other system theoreticians are not only interested in the way corporations react to environments, but are also concerned about the way information needed on the way is being processed (Terreberry, 1968, pp. 590-614). Interface functions are differentiated, attention is paid to non-market relationships (Brodén, 1976, pp. 46-49), relative corporate autonomy is seen to depend upon the type of environment (Rhenman, 1973, p. 190), and a different type of environment may require a different type of response. In contingency theory, attention is also paid to corporate structures that must fit the environment (Burns and Stalker, 1961, pp. 119-121, and Lawrence and Lorsch, 1967-1, p. 209, and 1967-2, pp. 1-47). The more a company differentiates its interface functions, the more it will also have to integrate all its sub-functions. Organizations may have an organic or mechanistic structure. Child (1972, pp. 2-23) observed that maybe there was no such thing as just one solution per type of problem. Others, such as Starbuck (1975, pp. 1069-1124) made it clear that the relationship between a corporation and its environment often is a two-way street. Sometimes large companies can even afford to have mechanistic structures in turbulent environments (Schryögg, 1978, p. 78 and pp. 282-283). Particularly interesting is Thompson's approach, his central theme of uncertainty as the essence of the administrative process in corporations (1967, p. 159), and his detailed analysis of "dependency relationships."

Management Theory

Ansoff's evolution of thinking since 1965 may be considered to represent developments of management theory at large. In a way these theories are variations on the organization-environment theme in which top-management tries to decrease its dependence and uncertainty by making strategic decisions, formulating objectives and how to achieve them.

At development stage one attention was focused on formulating strategy. This formulating approach then developed into a planning one, focusing on systematic procedures. Following the first oil crisis—and aware why strategic planning based on rational problem solving and decision making (Ansoff, Declerck and Hayes, 1974, p. 64) had failed—top-managements realized that it was also necessary to maintain good relationships with non-market stakeholders. Ansoff started to perceive rational planning processes as part of complex socio-dynamic, learning processes. He began to understand that in successful management systems, the human actor is just as important as decision making technology. So strategic planning developed into strategic management. As from the seventies, strategic surprises and paradoxical situations greatly increased. Increasingly companies were forced to develop an "after the fact responsiveness" (crisis management) or a "before the fact strategic preparedness" (Ansoff, 1975, p. 2). If companies are waiting to respond until the moment that a signal has become strong, the situation may have become critical. In order to prevent too early or too late a response, it is better to reformulate a response sequentially, and develop internal responsiveness accordingly.

Finally Ansoff also began to pay attention to the implementation side of policies, notably to strategic behavior. Over the years, corporate attention added the entrepreneurial sub-environment (integrating a longer term profit focus) to the competitive sub-environment (aimed at short term profits). After that, the political sub-environment was added, aimed at realizing societal legitimacy. Increasingly companies cannot do without a blend of responsivenesses anymore, such as market responsiveness, cost efficiency responsiveness, strategic responsiveness (when technological and economical discontinuities occur), structural responsiveness (see "structure follows strategy," Chandler, 1962), as well as internal socio-political responsiveness and societal responsiveness (Ansoff, 1974, p. 3). Policy can no longer be separated from implementation. In spite of all theoretical renewals over the years, Ansoff still put the accent on a normative approach, in which corporations should analyze problems and develop technical solutions and ("sophisticated") management systems "objectively."

Change, an Action-Sociological Approach

Silverman (1986, pp. 233-234 and 1970, p. 150) has been amongst the first not to approach organizational change one-sidedly. Organizations are systems of power, as well as value systems. His action oriented model clearly perceives that in each company role systems, patterns of interaction, and a number of (outdated) "rules of the game" have developed over the years. It is very important to know how actors are orienting themselves, how they perceive their situation and their goals, and how they expect others to see theirs. It is also of interest to know whether actions and interactions may lead to new situations, meanings, goals, means, the use of power and new outside interventions. Contrary to system approaches, in which actors are implicit factors or are supposed to function in pre-arranged ways aimed at normative goals, Silverman's action oriented thinking also, and particularly, tries to understand what has really happened in a situation and why actors behaved the way they did. What matters most in Silverman's approach is actors' behavior: (1) do actors (succeed to) change the rules, or (2) do they behave according to existing rules. Interactions between actors may follow from or lead to change. But how to proceed from this analysis model to a management model of change remains unanswered.

System theory assumes that corporations can only be managed effectively, if and when management has adequate information and an adequate set of management tools. Inspired by Dutch human resource sociologist Van Dijck (1979 and 1981), in situations of major change business goals are not normally formulated in an explicit and specified way. In those situations business managers have room to maneuver. Often they have different opinions about response goals and means. For effective management, managers must also have a managerial stimulus-response model at their disposal. In situations of major change this nearly always is an illusion.

In short, whenever explicit goals are lacking and interests and preferences vary widely, conflict is the rule and harmony the exception. Then rational management thinking can only explain decision making to a small extent. Then we cannot do without additional cultural and political paradigms.

Pettigrew gave power a central place in organizational decision making in situations of major change. He has shown (1973) why innovative decisions are far from rational in many ways. Not so much because of lacking information or an inability to "translate" information into new response policies, but particularly so because of power and interest relationships. In political arenas actors jointly seek new balances, with or against each other.

To a large extent organizations in situations of change are "political systems" not so much as dominant management coalitions, because then we would simplifyingly turn the situation into an "objective" problem, looking for "technical" alternatives before deciding. If we did just that, we would be looking for the normative views of the dominant actors without explicitly dealing with the decision making process, and without opening up the "black box." We are not following these lines of the technically objective variation of the political system approach (Easton, 1965 and Dye, 1972). What we need is an integrated model (see amongst others Jenkins, 1978 and Van Dijck, 1981, pp. 25-26). A model that actor orientation integrates the rational, "decision-logic" approach of most of the economically oriented organization and management theories with a socio-cultural and political approach. Such an actor oriented system approach especially pays attention to the dynamics inside the networks of actors. Some of the core elements of that blended approach are:

1. Top- and upper-level managements are managing decision making and learning relationships with other relevant outside and inside actors. Interactions greatly depend upon the specific power culture and power dynamics of a company.
2. In-company and external corporate response intentions and behavior take place in "multi-way streets," and follow from or lead to change inside and outside.
3. Policy decisions are important, but what matters most is what is being implemented and how and with what results. To which extent do these relationships determine the nature of the decision making process, and how can managers (learn to) better manage them.
4. Perception, vision, preference, choice and action are often strongly culturally interwoven with ideologies, values and beliefs, norms and attitudes. When company managers and outside actors express their views with regard to goals, means and results, at a close look one can pinpoint whether their respective interests are in harmony or in conflict.
5. When too many new problems arise and have to be tackled in too short a time, management may feel overtaxed in terms of learning, financing and mobilizing an organization. And may (instinctively) overreact by doing nothing.

Why Is Marketing Theory Not Actor Oriented?

Entrepreneurs and managers have been practicing action-sociology for ages, consciously as well as intuitively. That was not recognized by sociologists before the seventies however. Neither have organization studies shown much interest in these theoretical developments. Child (1973) noticed that the media have been "more alert to the political nature of organizational decision-making than many professional social writers on the subject" so far. In marketing theory and empirical studies the situation is not much different. We cannot leave marketing aside, because in addition to theory we will be studying how top and upper management levels of the national sales company of Philips in Holland and how the CEO top management level of the Group of Philips companies are learning to manage decision making processes more responsively and more interactively. We specially deal with marketing, because it plays a dominant role in these studied processes.

Apart from some interesting exceptions (see Turnbull and Cunningham, 1981, and Hakansson, 1982) in industrial marketing, nearly all theoretical marketing approaches follow a positivist, decision logical, rational and objectivistic line (one example out of many: Kotler, 1973, 82, and 1967 through 1988). "Traditional" theory in fact abstracts from actors who largely base their behavior upon "constructed meanings."

Second, "traditional" marketing theory very much thinks in a structural and functional way, and implies that "intellectualism" will mechanistically lead to implementation. The structural concepts, models and elements therefore remain empty typifications. They do not try to explain how to fill in the professional and organizational specifics to make things work. "Traditional" marketing theory tells students and managers that management must be responsive, must change strategies, programs and activities, or should purposefully decide just not to change. In that way theory tends to be normative and assumes that archetypical, ideal operational settings prevail. In situations of major change the contrary normally holds true. "Traditional" marketing theory does not therefore fit well in the innovative times of the day.

Thirdly, marketing academicians are not only doing research and publishing results, they are also teaching future marketing professionals. Unconsciously and unintendedly many of those teachers may have "learned" to identify themselves with marketing management practices of leading companies and are largely focusing on today's "hard" facts, actions and results, leaving actual managing and decision making, involving unforeseen and unfamiliar new (non-market) developments in changing social (that is: organizational) contexts, out of their paradigmatic scope.

And last but not least, however open many marketing practices in the outside world are, they do not easily let outsiders into their real-life, in-company marketing "kitchen," and do not let outsiders see and hear the marketing "cooks" at work in their own familiar surroundings for a long time. When we tried to get the cooperation of a number of major (multi)national corporations in Holland for our longitudinal, in-depth study, we met strong opposition there generally (Netherlands Employers Federation, 1978), and got a diplomatically phrased "no, thank you" answer from the country's leading retail chain in food (Albert Heijn, 1980). Following strenuous efforts, Philips (as well as Shell and a major bank) fortunately were willing to participate.

Business Ethics – From Rhetoric to Action

This treatise particularly deals with situations of (major) change, especially focusing on environmentalist, consumerist and (geo-) political non-market changes. These types of changes probably are amongst the most difficult ones for management to tackle. Let us see what has been written about business ethics and social responsibility. In the context of this treatise we limit ourselves to a number of basics.

Many critical things have been said about corporate social responsibility in Holland in the sixties to early eighties. Most of this criticism was looking "outside-in." One should also deal with it "inside-out," however, and pinpoint management's perceptions and actions. Often the outcome was an emotional taboo sphere full of rhetoric and good looking intentions, monologues and unsatisfactory stalemates.

In order to make these things more culturally compatible and manageable for management, it has been quite useful that some writers (Dierkes and Bauer, 1973, p. XI, Bauer, 1974, p. 8A, and Ackerman and Bauer, 1976, p. 6) have introduced a more action oriented approach based upon "response" and "responsiveness." It helps when response is put into a perspective of maturing new issues, corporate role and developing responsiveness, realising that response processes are learning processes in time.

First, Bauer's model of short term response processes is made up of three sub-processes: identification ———> commitment ———> implementation. Management begins by searching and "finding" issues. Bauer assumes that issues are not taken seriously before they have adequately matured, however, he leaves the implied maturing process as a "black box." He also abstracts from issues that happen all of sudden and in hard force. On the other hand we could not agree more with Bauer, when stressing that response processes do not end before the implementation has been lastingly secured, including a firmly reformulated strategy and organizational adaptation. Characteristics of response processes may change in time. Complexity and relative importance of the subprocesses also largely depend upon the situation and issue concerned.

"Identification," "commitment" and "implementation" are largely matters of perceiving the corporation's role, interests, power relationships, and the issue. What is seen as a societal threat yesterday, may turn into a business opportunity tomorrow. "Commitment" is often thought to deal with deciding which issue to respond to or not, implying a well-considered, autonomous behavior. Mostly real life is quite different, however. The issues involved are complex and uncertain, and have to be tackled in interaction with others generally. Quite often it is not a question of to respond or not to respond, but of "how," "how much" and "when." Again: perceptions. Regarding "implementation" it is important to know who took the response initiative, and also at which level the response decision has to be implemented. Does a manager have to implement his own initiative and decision or somebody else's, and who gets the fame and the blame? All of these sub-processes are also "technical" and managerial learning processes: how to learn and use the available response potential of the organization better.

In the seventies, Ackerman (1973, pp. 88-98 and 1975) studied why major divisionalized corporations in the U.S. were having problems of integrating societal issues into corporate policies. Ackerman was particularly interested in "responsiveness," how the capacity and willingness to respond may change and develop in the longer term, and how the organization of the response implementation may change in time. Organizationally, feeling engaged tends to follow a three step sequence. First, top-management begins to air policy intentions. Second, a staff specialist is appointed to produce new types of information for top-management. During these first two steps, actor learning dominates. Third, the issue increasingly changes from a societal problem into a management problem. One begins to understand that it is an interwoven problem of policies, information processing, motivating people and management systems on the one hand, and implementing response on the other. Contrary to "technical" learning, the accent is on "administrative" learning now (Murray 1974, pp. 6 and 12-17, and 1976, pp. 5-15). In our approach responsiveness is the way in which management learns to create, incite, organize and implement the kind of responsiveness needed. In our dynamic learning and decision making model, responsiveness is one of the major dimensions. A dimension in which our modification of Ackerman's steps is blended with Ackoff's attitude oriented response typologies of inactivism, reactivism, preactivism and interactivism, the last type representing a situation in which a corporation sets its own future (1974, pp. 22-23).

Our responsiveness spectrum ranges from conservatism via testimony and analysis to implementation, ranging from (un)consciously not willing and able to change that willingness and capacity on the one hand, to consciously responding reactively or interactively in ways compatible or incompatible to the wishes of others on the other hand.

The dynamic core variables "information," "culture" and "power" have different meanings, as per type of responsiveness.

1. Information. In the conservative situation management is unaware of the societal development concerned, because they may be weak signals, or may be negative ones for the company. In the testimony situation developments become "harder" and more specific and one cannot any longer deny them. Threats come closer quickly. In the analysis situation management begins to analyze the issue's relevance, the response alternatives, the profit or loss situation and manageability of the developments. And in the implementation situation, management has a clear picture of all relevant dimensions.

2. Culture. A conservative management hardly feels itself societally engaged. Short run self-interest prevails. In the testimony situation management states that it wants to become more responsive. Management intentions are aimed at stimulating the company's societal intelligence, if not responsibility, and at defusing outside criticism. In the analysis phase, management finds it necessary and assures that the situation is properly analyzed. Reaching the implementation situation, management is prepared to choose and implement a specific response. Intentions: pre-empting regulation threats, or gaining a political or competitive edge.

3. Power. In the conservative situation, developments cannot force management; outside actors have to accept an unsatisfactory situation. In the testimony situation, management manages a changing situation verbally. In the analysis situation, management keeps close contact with purposefully appointed consulting specialists. Contacts with other managers and outside actors increasingly become negotiating relationships. And in the last and implementation situation, the required new or re-activated organization to implement the chosen response and to manage the implementation, is put in action.

A Dynamic Model of Learning and Decision-making

By combining our above theoretical considerations, we have in fact put together a dynamic model of learning and decision-making. Simplifyingly this model can be shown in Figure 5.1.

For empirical research purposes a few more things on "culture" and "power" have to be added. (By culture we mean the dominant sets of shared or split values and beliefs that set the norms, rituals, heroes and symbols of behavior. And of behavior itself. And of the ways management is supposed to do things, to choose, to interact and to make use of power potentials.) One should not be surprised to find that in many a company, culture is not (yet) an explicit part of management's functioning. Mostly because of cultural incompatibility—culture being considered unfamiliar, "soft," not relevant. Managements frequently take the company at large, its goals, structures, and procedures, for granted, as if they had objectively come into being in a natural way. The company's staff often accepts them as the truth. Even if management does not mean it that way, they often "monopolize" and set the way in which the company's realities are constructed, continuing a fiction of rationality. Studying corporate (power) culture is a difficult affair. If a company would welcome an assessment of its crucial values, mild conflicts might

FIGURE 5.1. A Dynamic Model of Learning and Decision Making

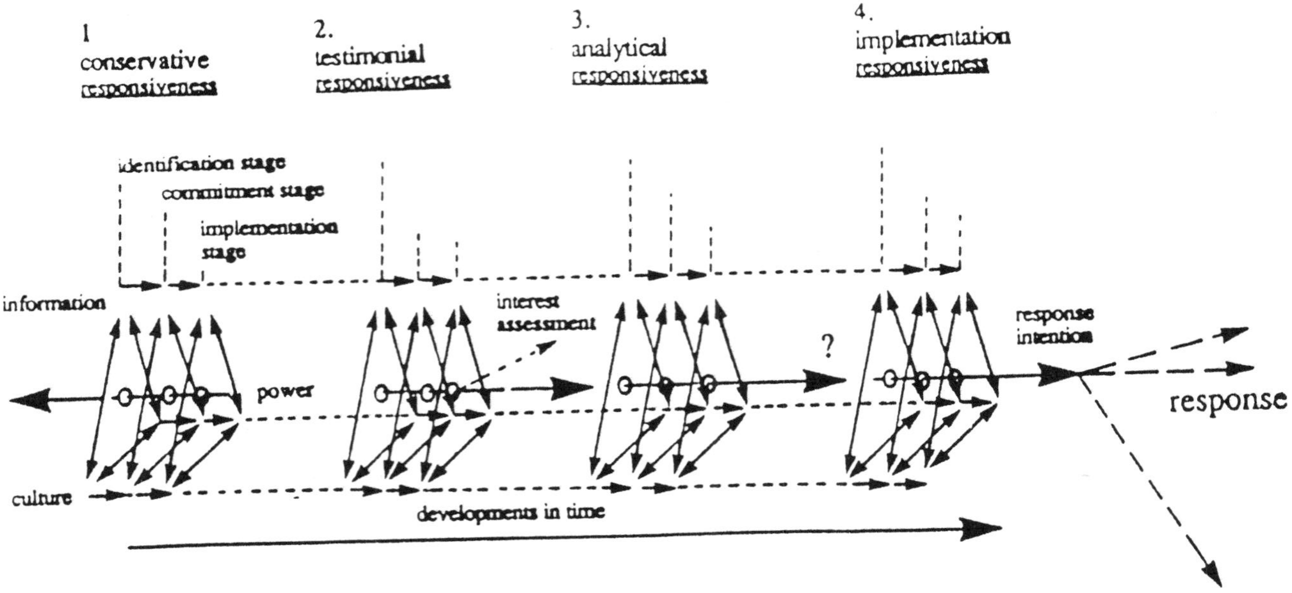

develop into explosive ones. Chances are that some of the dominant actors might loose some of their hierarchical, professional or charismatic power, or may not be able to use it as efficiently and effectively as they used to. Quite often organizational values and beliefs, as well as power relationships, will only be made explicit, as long as it serves the dominant actors well regarding their power relationships and the organizational manageability they are after.

Language is a cultural phenomenon. Management creates and uses language in many situations for many purposes and in different forms—in writing and in spoken words, openly or privately, officially or off the record. As Mill once said (1972, p. 62): "A vocabulary is not merely a string of words; immanent within it are societal textures—institutional and political coordinates. Back of a vocabulary lie sets of collective action." Language is communicating action and is part of the action. Language is the most important tool of analysis for our empirical study inside Philips. Language has been carefully studied during many long and short talks, meetings, conferences, and observations of some 25 Philips managers. In addition, written language was studied, ranging from confidential internal documents to official corporate public statements.

PHILIPS CEO CULTURES AT THE MULTINATIONAL TOP LEVEL

Cultural Constancies

First let us look at the constant dimensions. Corporate thinking and doing have been based upon a number of fundamental corporate values, such as the competitive system and internal hierarchies. To secure superiority and success, the company must be free to choose the technologies and products it needs for its leadership position. Philips feels proud to be technologically respected all over the world. Not Holland, not even Europe, but the world at large is the corporation's home market. Aggressive and timely marketing have been underestimated for a long time. "Wait-and-see" attitudes have prevailed too much, following a preparedness strategy letting others mobilize new markets first.

Time and again, Philips tells the outside world that the company "is doing big things, thanks to her sharp and visionary entrepreneurial eyes" (Loupart, 1958, pp. 237-238), saying that she is serving "real needs of the people everywhere" (van Riemsdijk, 1974-2, p. 8). Product breakthroughs almost always get ethical justification: black and white and color TV, video recorders and compact discs were "enriching and developing the people and were bringing jobs to tens of thousands of workers."

Intra-company discussions about new products were strongly technically, economically and functionally flavored. Consumers were believed to know what they want, of course Philips was answering their wishes, they said. What society should look like, on the other hand, is the politician's job, not Philips'. Consumerism may have had some awakening effects, but in most cases it was not a real opportunity or threat for the company. Over the years the chief executive board has learned to "separate society's chaff from wheat." In her view, Consumer Unions in Holland only reached relatively few and biased intellectual consumers, who are not normally buying the way the unions tell them they should (van Riemsdijk, 1973, pp. 2-3).

In responding to increased criticism, the company often tried to silence critics, or played a professional "one-upmanship." "Critics—(van Riemsdijk, 1974-1, p. 1) are not well informed, they are not discussing the way they should, or make a suggestive, emotional and rather black and white impression, and do not always seem to base themselves upon a careful analysis." At times critics are called "dogmatic" (van Riemsdijk, 1974-2, p. 2), are blamed for "mixing up things ignorantly or wilfully" (van den Brand, 1980), or are said to "utter opportunistic political cries" (Otten, 1980). According to 1981 perceptions of the chief executive board (Otten, 1981) ". . . many critics have such a comfortable place outside business, that they can do away with the facts and that they can allow themselves a popular stance in all freedom of conscience."

Philips culture was strongly based upon self-interest. The company's role of the old days of "producing goods and services of desired quality at competitive prices" has changed into today's role of "creating a climate, in which the company can realise its goals"

(Otten, 1981). The core of the matter remains her well-understood self-interest.

It should be emphasized, however, that in spite of the above testimony of responsiveness, major marketing efforts of the company are not aimed at creating a better understanding, but at securing satisfactory sales results instead, including strengthening market positions and better profits. However much the chief executive board may testify that the company is ". . . self-evidently developing and presenting new products without forgetting market and societal acceptance" (Annual Report, 1978, Otten, 1981 and Dekker, 1982), at the same time the company keeps on refraining from explaining its energy stand to the public. Even in 1979 it is still exceptional for the company to put its promotional accent in lighting advertising on energy friendly products. Business as usual is still the order of the day. Doing so, by hiding herself behind governmental authorities (van der Klugt, 1980): "Many technological innovations depend upon the preparedness of the central government to create the necessary infrastructure. By doing so, the government at the same time accepts the moral responsibility for that kind of innovation."

For a long time many members of the chief executive board were not aware that bureaucratic characteristics made the organization conservative and slow. Additionally, remnants of patriarchal and dualistic family relationships of the old times, stemming from the technical and commercial founders of the firm, had strongly stimulated paralysis. Too often top level management was thinking in terms of "problems" and too little in terms of opportunities and challenges. Criticism was seen as threatening. Non-change was normatively preferred to change. Small changes were considered better than big ones. And in spite of the corporation's near-global nature, successive chief executive boards were fully manned by Dutchmen.

Cultural Variables

In addition to constant prime interest for economic themes, attention to new environmental issues is shifting in time. In the early sixties the company's "Fundamental Views" (Loupart, 1958, pp.

224-238) pictured an ideal growth company serving customers, and guaranteeing "a happy and prosperous life" to all people employed by the Philips' democratic federation of companies.

In the late sixties and early seventies continuity thinking was based upon "exchange" (Noordhof, 1970, pp. 9-39) between society setting favorable terms for industry and corporations accepting social responsibility as long as profitability was not unacceptably affected.

The chief executive board (intentionally) understood that the corporation needed more responsiveness which could only be achieved when managers could be taught less resistance to change, and when they were given a clear picture of all major corporate developments. Openness should increase and responsibilities should be more decentralized. The CEO also knew that harmonious ways of cooperation can only be stimulated, if motivation and participation can be improved. These views and intentions do not, however, automatically mean that the changes wanted would indeed take place. Less so, because Philips managers in the many operating companies often had varying opinions, intentions and routines of their own.

As the "outside world kept on criticizing unjustifyingly" (van Riemsdijk, 1973, p. 2), as from 1973 the CEO thought that the time had come for Philips to begin to justify herself to the outside world openly by defending private enterprise to be most appropriate to respond "timely and effectively" (Annual Report, 1973, p. 5) to society's needs and wants, and to produce the things "consumers ask for" (van Riemsdijk, 1973, p. 3), "through market mechanisms" (van Riemsdijk, 1973, p. 6).

At the same time the chief executive board emphasized that many interests of stakeholders are contradictory. Philips had started to "integrate (some of the) new societal developments in her Research & Developments," such as improved energy efficiency in lighting (Annual Report, 1973, pp. 5-6). Statements like that are suggestive ones because, in fact, marketing practices actually went on to please consumers who wanted more lighting, not less energy consumption (Annual Report, 1968). Regarding domestic appliances, marketing policies were not really affected either. These practices also remained on their normal sales oriented track. The continuing economic primacy can also be seen from the Board's approach to the first Arab oil crisis of 1973/74, translating this "in terms of less growth, higher costs and prices, and international monetary implications" (Annual Report, 1974, p. 5).

Amidst continuing strong criticism of multinationals in 1974, the CEO chose sides on fourteen different power related topics (van Riemsdijk, 1974-1 and 1974-2). One of the criticisms regarded the (supposed) manufacturing of societally unwanted products. In the list of the fourteen topics, the issue of "unwanted" products took only a low eleventh rank, denying corporate responsibility, ". . . for it is the consumer who decides how to spend his money in the last place."

Further to the "Starting Points of Corporate Policy" of the 1969-72 CEO, successive "Basic Thoughts on Corporate Policy" came into being in 1975. As before, continuity ranked first. For the benefit of all parties involved. These "basic thoughts" included eleven "starting points." Technological push remained highly important for the company; employees and investors were prime stakeholders; and compared with the past the consumer had moved up to fifth place on the list. This indicates that societal conscience and intelligence of the chief executive board were growing. On the other hand, direct market responsiveness remained of prime concern (see statements like "Philips produces what the consumer wants and pays much attention to the usefulness, robustness and safety of its products").

In the mid-seventies the CEO became aware that corporate strategy could only be successful when two additional conditions would be met: (1) more attention to the importance of longer run objectives, technological innovation and, a better organization and manager training about what is going on in the world, and how to respond to developments adequately, and (2) management should think more about the corporate role and should formulate a future vision. Top level specialists were asked to have another look at "strategic planning, external relations policy and organizational structures" (van Riemsdijk, 1975, p. 12). These intentions clearly showed a shift from pre-1975 days, when "it was the usual habit of the chief executive board to do what everybody else was already doing" (Frits Philips, 1976).

This evolution was confirmed in 1976 when for the first time in modern history Philips explicitly took societal sides (Annual Report, 1976, p. 8) stimulating society to formulate her opinion about the role and practices of responsible businesses and adding response intentions such as: Philips should actively participate in these societal developments, but should keep wanted changes out, if society would be overasking. The company would not have her decision making autonomy endangered.

After a new CEO joined the "Board" in 1976, culture became markedly different again. Increasingly, the company perceived herself as the weaker, dependent actor, able only to adapt and at the same speed as others (Rodenburg, 1977, pp. 3 and 6-7). The company was only prepared to consider societal "conditions," as long as they were not "contradictory" (Rodenburg, 1978-1). Because of severe international economic adjustment processes, and a far more critical environment, Philips found it more difficult to realize its goals and to contribute to "a maximally useful state of employment" (Rodenburg, 1978-2, pp. 58-59). However much society considered multinationals increasingly "powerful and rich," the chief executive board was experiencing "a corporate profit squeeze, and a growing feeling of weakness" (Rodenburg, 1978-2, pp. 68-69). Now consumer legislation was to be reckoned with by Philips (see a member of the chief executive board stating that the company "should and does loyally adapt" to that kind of legislation, [Dekker, 1978]).

In 1979—one hundred years after Edison—the CEO explained that "energy efficiency has always been a criterion for developing new lamps. Increasing light output in comparison to energy input has always been important" (Rodenburg, 1979, p. 13). For decades marketing strategies had aimed at greater light outputs at equal energy inputs. The "equal amount of light output at lower energy input" variant only hesitatingly entered new product development thinking in the seventies. That development was not market or marketing pulled, however, but primarily technologically pushed.

Although new product development policies for consumer lighting products started to become more energy friendly since the first oil crisis, the chief executive board did not initiate or approve more responsive marketing operations for consumer products generally. That is why the 1979 CEO said (Rodenburg, 1979, p. 19) that he did not want to touch the subject of Philips' market activities. The company remained market oriented, as can also be seen from a statement of the chief executive board (Annual Report, 1979) on the electrification potential of private homes: ". . . there are still opportunities for new products meeting new functions." In other words, there were still lots of unsatisfied electrical needs and wants for Philips to satisfy.

In the energy issue Philips took self-justifying stands. In 1973 the company internally even denied and refused to accept any energy responsibility in marketing (Philips Lighting News, early 1974). The arguments of West German professor H.W. Bodmann were explicitly endorsed, who warned against false savings in lighting and pinpointed cars and central heating—in which Philips did not have a stake—as the real saving potentials (Frankfurter Allgemeine Zeitung, December 28, 1973).

It should be noted that, in the U.S., General Electric responded strongly in 1973/74 by implementing its Energy Conservation Council, making sure that GE would use 10% less energy as soon as possible, "to keep GE people at work and to continue serving customers without interruption," and by aiming its advertising efforts at GE's most energy efficient products (Advertising Age, February 4, 1974).

Even in 1979 the Philips CEO was still shifting the blame on energy unfriendliness in private homes to others, notably to central heating (Rodenburg, 1979, p. 13).

In fact, Philips' strategic choices on energy were unclear and contradictory in the seventies. The Board was split into three opinions. Strongly advocating energy friendliness was the board member representing Research. A mixed pro- and con-group corresponded with the manufacturing divisions. The largely "con" third group came from the national sales companies of the corporation. As long as the energy discussion in the Board went on, without clearcut strategic choices, responsibilities for choosing more or less energy friendly policies were left to the discretion of the corporate research organization, manufacturing divisions and national sales companies.

One of the board members—notably the one who became CEO in 1982—sounded much more responsive. In 1978 he advocated higher priority to newer societal developments, such as ". . . the environment, energy, consumerism, advertising ethics, codes of behavior and corporate legitimacy" (Dekker, 1978). And one year later Dekker made it clear (Dekker, 1979-2) that another type of manager was needed to successfully manage R & D efforts and outside criticism. That new type of manager should also be able to manage and communicate policies consistently. Managers should be qualified to have good relations with government authorities and the consumer. Through timely consultation and dialogue they should determine how much and how soon society wanted its innovations.

During the late seventies and early eighties, these newer tendencies kept on developing in innovative and no-nonsense ways. Two Board members made that particularly clear. They believed in themselves and called a spade a spade. They explained that for a company like Philips, operating in a societally responsible way is far from easy. They let it be known that organizational steps were underway, but that many more new ideas and stakeholders' interests were still waiting to be integrated into the company's planning and decision making apparatus, both at multinational as well as national levels. Philips—according to Dekker (1979-2)—should formulate and communicate understandable and defensible standpoints, defensively and offensively.

In 1980 another member of the chief executive board made it clear that strategic choice for Philips meant that "it will be impossible to assure a social paradise built on economic ruins" (Kuilman, 1980). From 1982 the company's articles do not state any longer that Philips aims at "a maximally useful state of employment" but that Philips "can only accept and realise her social responsibility, as long as she remains financially healthy."

Cultural Summary

Three cultural periods can be distinguished during the sixties to early eighties.

The Sixties

These are years of a growing economy, sales and profits: years in which the company profited from rising purchasing power and market opportunities. The client was king. The company's attitudes were mainly short run. Technology was her driving force and everybody's pride.

Employees lived in a "social paradise." Criticism was hardly accepted. The corporation felt infallible, self-complacent and haughty. Successes resulted from Philips' strength and failures from bad luck or from the outside world.

The Seventies

The economic situation was changing. Markets began to saturate. Japanese competition arrived. Unwanted, societal side effects appeared. "Profit" got a negative ethical meaning. The word "multinational" became loaded. The company felt the aftermath of student revolts in places like Paris and Amsterdam in 1968. The first oil crisis "broke out." Developments were perceived as threats and these were approached in escapist ways, often creating unreal excuses and solutions diplomatically or normatively packed, denied or nullified, or counter-attacked by "borrowing" arguments from sympathizing independent experts, or by exchanging monologues.

Testimonial responses and intentional behavior were suggestively worded as if they had already been implemented, repeating the same phrases all the time, and making them into rituals.

Longer run thinking and planning were appearing. The company's big computer venture became a flop, bringing its myth of infallibility to an end. The chief executive board became more socially and societally intelligent. Conservative responsiveness tended to change into testimony responsiveness, sometimes even into analysis responsiveness, e.g., by studying the possibilities of setting up a modern managers' training course (Annual Report, 1976, p. 18).

After 1976 the economic times became tougher with recession and global Japanese competition. Options could no longer be kept open all of the time. Priorities had to be set. Technological innovation seemed to come to a standstill, (too) few new products were introduced. Dilemmas were all around. Frequently better product quality and longer product life were raw material unfriendly. Individual consumers quite often did not behave the way they "should," if they were to follow consumerist "preaches." Internally, productivity surpassed market growth, staff had to be fired and the "social paradise" was lost underway. Philips was demythologized and became the national scapegoat of the unions in Holland.

All this created a reserved, cautious, defensive and puzzled attitude (van Hamersveld, 1977): "How can consumers determine which products should and which ones should not be marketed? How can one judge a product that is societally useful on the one side, but uses a great deal of energy on the other? Which criteria should one use for product choice now vs. later? And how does one manage all that in an economy as internationally open as the Dutch one?"

The seventies were also the years in which the chief executive board made itself invisible for most of its operating divisions and companies. The company lost its grip on the media, leaving Philips to decide whether or not to react to bad publicity. And when it reacted, it often did so in an uncertain and uninspired way, making pessimistic and compromising policies based upon hard internal bargaining and disagreement, felt. Those were the things that set the company's morale of the times.

Late Seventies and Early Eighties

It became clear that the more realistic and optimistic members of the chief executive board were getting the upper hand. Outside and inside inner circles saw the first signals of a more direct, pragmatic, open and "political" approach by the company. Philips' top echelon was becoming more aware of dilemmas and the importance of relationships. Readiness to enter dialogues was increasing. These stronger board members were also showing that they were more action oriented and less focused on company weaknesses and outside threats—doing away with the problem trauma years of the seventies—but sparking off a new inspiration, and a new belief in one's own abilities. They sought and created new opportunities, and began to do business in a well-considered way and at a higher speed.

In these 1978-1982 years most managers at operating companies may not, however, have been able to see this emerging new management style and vigor. As from 1982, when Dekker became CEO, these changes got clearer every day and showed that Philips was beginning to consider herself "a quasi-public body, increasingly being called to account for her "hows" and "whys," and unable to turn away from political processes in legislation and regulation. Dekker made this quite clear in a lecture during a closed meeting with ambassadors on January 22, 1981.[1]

As the new CEO, Dekker made it widely known in 1982 what his turnaround intentions were: "It becomes increasingly difficult to separate business from politics. The enterprise cannot avoid—at least—to be conscious of politics and to form an opinion about them. The CEO of an enterprise must wield such a motivating influence on his environment, on all corporate layers, as well as outside of the company, that Philips secures her position of confidence and trust, that she is and will be a company of success. Early to rise and hard work, because otherwise you miss the boat."

MANAGEMENT CULTURES OF THE NATIONAL SALES COMPANY IN HOLLAND

Introduction

Having examined Philips' CEO cultures at the multinational top level, let us now look at management cultures at the level of the national sales company (NSC) of Philips in Holland.

In 1980 the Philips multinational group of companies—in addition to staff departments—was made up of a Central Research Group, fourteen major Production Groups, and sixty-four National Operating (Sales) Companies. In consultation with the Production Groups, each NSC integrates its national policy into the country in which it operates. The Production Groups are setting their international product policies in consultation with the major NSCs. The degree of influence of an NSC largely depends upon the nature and importance of its market, the type of product and the organization of the NSC.

Total staff of the Group in 1980 numbered over 370,000, of which some 80,000 were in Holland (Dutch NSC: 4,000). It may not be surprising to learn that Holland is one of Philips' key countries. Although its relative importance is decreasing, in 1980 it (still) made good some 9% of total global sales of the Group. Around 50% of total sales regarded consumer products, leaving the balance of products for industrial markets (in Philips' jargon: "professional products"). In the NSC, consumer products are split into a number of business units, notably (major and small) Domestic Appliances (including products for personal care, lighting and dry batteries), Audio and Video.

In this section we deal with the cultures of the general management of the NSC and the management of its business unit "Domestic Appliances." Contrary to the Group's level, at the NSC level response intentions and behavior resulting from complex interactive processes in time—in short, the decision making dynamics—will be analyzed in greater detail. We will look into three types of cultural sub-relationships, however hard to separate from one another empirically. These three sub-relationships are "informational," "ideological" and "political," as explained and formulated in the theoretical, dynamic model of the first part of this exposé.

Informational Sub-relationships

What does the general management level know? What do they know themselves, about each other, about their relevant Production Groups, about the CEO and the Chief Executive Board, about the outside world, about power relations, motives, intentions and behavior, about tightness and looseness of systems and procedures (e.g., the way management guidelines regarding sales and profits go "top-down," and how plans and budget recommendations go "bottom-up," including the inflating and deflating that goes on during the processes)?

All of the NSC and business unit managers have a long company history, ranging from ten to twenty years. They served the company in different product, marketing or general management roles. The NSC's CEO is the only general manager of the NSC who can boast international management experience. All but one are Dutch nationals.

Company training courses have taught the managers that Philips goals are "profitable sales and satisfactory returns on investment, to be realised by identifying, anticipating and satisfying user's needs and desires" (van Rees, year unknown, p. 6). From these courses participants also learned that management should concentrate on "assessing and converting customer purchasing power into effective demand for a specific product or service" (van Rees, year unknown, p. 7), and that managers must be societally conscious and able to determine how non-market stakeholders will influence market processes sooner or later (van der Klooster, year unknown, p. 17). Managers are also taught that consumer interests are not more important than the interests of Philips: ". . . serving customers' interests cannot take precedence above all else; benefits for both sides are aimed at; i.e., satisfaction for customers and profits for the company . . ." (van der Klooster, year unknown, p. 20). These courses do not, however, teach managers how to handle new types of response successfully.

In management's daily practice the consumer is mostly a far away abstract "thing." On the other hand, peers, the boss and the boss' boss are only x-doors or y-floors away. Consumer interests and satisfaction are "soft," complex and subjective, whereas the interests of the company—sales and profits in terms of money—are "hard" and "objective."

Until 1956 management of the NSC knew all the answers. Asking questions of consumers was a waste of time and money, not even permitted. Until 1966, management did not allow the company to have contacts with consumerists. First contacts with governmental authorities and the Consumer Union date back to 1970. Managing external relationships systematically did not start before 1978, when a new CEO entered the NSC.

As for market research, by the early eighties management had come into a state of crisis, because in the sixties and seventies they had told the world: "we are researching everything, so we have answers to all problems." In the meantime, however, the very same management learned the hard way, that management technologies were in fact still in their infancy; as to planning the future: "It is a hell of a job to tune future products to future demand to-day." The results: consumerist criticism and lack of management know-how to turn around market threats.

As from the late sixties, management responsibility was delegated by NSC's top-level management to business unit management. Although NSC management courses started in 1961 (at the initiative of the then NSC CEO), a marked professionalization of business unit management did not begin before the seventies. In the meantime management problems and responsibilities outgrew the improved management capabilities learned on the way. Some managers realized, that the training courses made for "programmed" instead of creative managers.

NSC's management also was an important information gatekeeper. In the seventies it categorically blocked information on consumerist developments sent by the Consumer Affairs Staff Department at Headquarter from reaching NSC's business units, Self-justifyingly explaining that the multinational's Chief Executive Board and its consumer staff specialists stood too far away from reality, and that management of the business units was too close to the market. This so-called "tune-fork thinking" meant that if NSC's management would vibrate too much, business unit managers would vibrate even more and might act very responsively, by doing so running the risk of not selling anything anymore. Sales attitudes that dominate at NSC's management level: ". . . if we have no choice, we rather suffer image than sales . . ." One should immediately add that with the arrival of a new CEO in 1978, the number of societally oriented top-down impulses in the NSC increased, even if the stronger implementation responsiveness that went with it was largely an incidental one.

Because of the new CEO, the style of management changed considerably. The closed, authoritarian type of decision making of the pre-1978 days changed into a more open functioning, in which NSC's management listened to arguments of others first, before drawing its own conclusions. That did not make the arguments—or information generally—more objective as such. As before, and because of increased complexity and uncertainty, the relative importance of "arguing by feeling," even tended to increase. Arguments became more subjective, but were "packed" in a more professional and objective way.

Managers knew that they were being judged from time to time. Every manager knew the rules of this game. He knew he was being judged for his management knowledge, his policies, his action orientedness, his initiatives, his company loyalty, his way of communicating and participating, motivating and manipulating. He was being judged for sales and profit results and how he did it. Knowing that everybody was continuously judging everybody else, managers behaved like "gallery players." Responsiveness aimed at new market and societal signals was not a part of the judging rituals (yet).

During the sixties to early eighties, not only at the multinational management level, but also at NSC management levels, new phases of responsiveness developed alongside and in addition to existing ones. Conservative responsiveness tended to develop into testimony responsiveness, into analysis responsiveness, and even into some occasional implementation responsiveness.

During the greater part of the seventies, NSC's management perceived consumerist problems mainly as "technical" ones of information processing, i.e., as problems that can be solved by arranging for new staff functions, such as:

1. "Consumer Interests"—stemming from Technical Servicing—handling individual consumer complaints and giving information to individual consumers. This department was following a somewhat hesitating and reluctant policy. It felt that an active policy might lead to far greater numbers of complaints and costs to handle them, without higher sales to compensate for them.
2. "Consumer Affairs"—aimed at controlling and stimulating technical product quality. The Consumer Unions were doing comparative product research." Consumer Affairs, was the technical liaison between the NSC and the Unions in this respect. By the end of the seventies "Consumer Affairs" had become so much societally engaged, that it recommended a new approach and organizational set-up for quality control of NSC's marketing policies and activities inspired by a publication in the U.S. (Divita, Summer 1978)
3. "Consumer Issues"—starting in 1970 as a staff function to NSC's general management. As of 1978 this job was taken over by one of NSC's general managers amongst others to intensify responses to external consumerist activities and politics. As consumerist issues matured politically, "Consumer Issues" increasingly intervened whenever necessary and possible. At the NSC, general management was also well informed about the undecisiveness of the Chief Executive Board of the multinational parent company regarding the energy is-

sue in marketing policies, for example, virtually leaving this subject to the discretion of the operating companies.

In the late seventies and early eighties NSC's top-management from time to time told outside professionals and academicians how the NSC thought about new product development, consumer research, societal responsibility and corporate strategy (Kuijpers, 1979, pp. 20-25 and Tuyt, 1981). Quite often the business unit's management was not informed about these public statements, nor were they well informed about new important consumerist developments, such as new government policy intentions and the new Dutch law on "Misleading Advertising."

As to energy, both management levels were equally informed, at least they perceived the issue identically, and in the same way as ad headquarters, i.e., in a rather dynamically conservative and self-justifying way. They assumed that consumers were not prepared to pay more for better energy efficient products, and that most domestic appliances were not energy unfriendly anyway. One was assuming a kind of consumer behavior that fit Philips best. NSC's management added one more dilemma to the justification by saying that however much small domestic appliances may be relatively energy friendly, some of the company's major domestic appliances were not, leading to the question of where to find the required large sums of money at short notice to develop new, energy friendly, major appliances, also bearing in mind that video developments had already cost large amounts of money and with the understanding that for most product groups R & D plans were fixed for years. As for major appliances, these plans aimed at more comfortable products at lower prices, not at energy.

To summarize, the informational relationships are:

1. NSC's management was well aware of the strategic thinking of the multinational's Chief Executive Board. She was also aware of all consumerist policy recommendations of the multinational's staff specialists, and of consumerist developments at large. She also knew what was going on in the Production Groups and in the important National Sales Companies abroad.

2. The business unit's management, generally speaking, was not (adequately) aware of the strategic thoughts, feelings, intentions and choices of the multinational's Chief Executive Board. NSC's management was purposefully keeping consumerist recommendations from multinational sources away from them. NSC's management did not compensate this blocking by given guidelines of her own instead. The business unit could only find out what NSC's responsiveness slack was, by ad hoc tests.
3. Vice versa, NSC's management generally speaking only had limited knowledge about the willingness and capacity of the business unit to act responsively.
4. The business unit was only taking consumerism seriously when signals were "hard," for example, when comparative product tests by outsiders showed negative results. The business unit also had learned, however, that good test results seldom lead to higher, and that bad test results seldom lead to lower, sales results for products of the well-established Philips brand, apart from unforeseeably sensational events.
5. To a large extent the business unit's attitude showed that management had great trust in its own staff specialists and external consultants. They did their job well and paid adequate attention to relevant changes. By this rationalization management could afford to keep its own job as little complicated as possible, and could shift societal responsibility to others.
6. Power was so fundamental and self-evident in all of management's actions and interactions, both externally as well as internally, that managers mostly went about it in a natural, implicit, and intuitive way. So much so, that they did not recognize and accept it when told. Often managers had a partial and biased power picture, which did not make decision making as responsive as it could (and should) have been.

Ideological Sub-relationships

The business unit of "Domestic Appliances" ideologically was a duoform one. The major domestic appliances' part was more functional and formal, marketing plans were only good ones, if in a "hard" way "proven" by impressive, written reports. In the group of the small domestic appliances, intuition and informal ways of doing things made much more sense. These ideological differences fit the consumer behavior that was relevant for each of the two groups.

The business unit cannot be fully appreciated without understanding the way the NSC at large and the rest of Philips were "doing things." There were quite a few situations in which technical ideology dominated, where products were largely used for functional reasons, where organizational attitudes were based upon economically rational consumers. That also made it understandable why the technical product was the number one actor in Philips advertising, why Philips advertising was advertising by a "proud engineer." Born out of technology, during the sixties and seventies, the company's technological developments – not her fundamental research – made Philips an imitator instead of the pioneer she once was, however. It became part of her active ideology to respond to discontinuities in defensive ways, i.e., when somebody else's new developments were beginning to show results, Philips quickly joined the party.

As long as it suited their personal or business goals, management did not have a moral problem to mind her own business and to let others mind theirs. But as soon as those goals came under fire, a double moral standard was used. Since its founding in 1891, Philips had found it quite normal to be free to build consumer acceptance for new electrical products and to stimulate consumers to buy them. As from the mid-seventies the Netherlands Consumer Union found certain types of electrical products quite senseless and strongly advised against them. Applying a double moral standard, a member of NSC's management team in a seemingly objective way called this consumerist behavior "undoubtedly unpermitted."

Over the years, the little light bulb factory had developed into a diversified, major multinational in electronics. Along that development Philips became a "dignified" company, also behaving that way publicly. Advertising was one of the things to be "dignified," assuring "respectful relationships" (Philips General Advertising Department, 1976) with customers. It should show "Philips the Great," a technical and businesslike – but above all – civilized corporation, seldom emotional and human. That lead to cautious advertising, quiet and serious, but also dull, colorless and faceless. As one NSC manager said: "Philips felt embarrassed to be in the business of selling."

These feelings of technical superiority also lead to self-complacency and haughtiness. In the early seventies the business unit's management said for example: ". . . the light bulb trade simply cannot do without us . . ." In policy making and trade relationships, the company behaved accordingly. So the NSC did no business with the up and coming, "unprofessional and banal" new channels of distribution. If it were not for a costly catching up operation in the late seventies, the company would have missed out on the larger part of the retail potential in light bulbs by that time altogether.

So professionalism developed into self-confidence first, and into self-complacency later. Over the years, outside criticism continued and developed into threats, perceived to be real. Over the years consumerists succeeded in building up momentum and inspired the government to plan and implement a number of consumer laws and regulations. This created feelings of greater uncertainty at the two management levels. Based upon the perception that everything a multinational said was suspect right from the start, this new uncertainty did not make managers speak up to the outside world. Sales and profits had to be cautiously defended. Managers felt scared that an energy friendlier positioning of company, brand and products would cost money. Or would trigger off even stronger criticism aimed at other product ranges, that could not yet be redesigned, leading the company from bad to worse.

This lack of self-confidence also had (other) inner-directed implications. Remember the blocking of consumerist information by NSC's management that indicated that management did not believe itself to be capable of managing the business unit's responsiveness and responses. In addition, business unit managers increasingly felt their self-confidence undermined because fewer new products came out of the R & D pipelines, and because TV and the press became very negative on Philips in the seventies generally.

Too long there had also been (too) much breeding in. Managers were rotating inside the NSC and the Holland based Production Groups, that is to say, Dutch managers were rotating in Dutch cultural settings. Some managers occasionally had international contacts, oh yes. But always as a part of their Holland based major Dutch jobs. At the NSC level, some managers had been on an international rotation scheme, and had the opportunity to check their Dutch oriented values and beliefs. It is also not without ideological meaning that the NSC—just as its multinational's headoffice—is located in one of Holland's Southern provinces, that is to say, outside of the more cosmopolitan Western part of the country. A Production Group with which the business unit had many contacts—notably Small Domestic Appliances and Shavers—is located in the Northern provinces, where—as one manager put it—"things haven't happened yet, and where the environment is too conservative to be stimulating." Bounded job rotations like these had some "provincial" effects upon the values and beliefs of the National Sales Company of the multinational corporation.

Over the years the decision making ideology of the NSC and the business unit changed quite a bit. That is to say, it became increasingly more interactive. Much more than before it took place in networks of actors. Managers were consulting each other more intensively. Decision making had become more of a joint affair. This networking had started from the structural division of labor between production and sales, and from the ever increasing financial scope and complexity of operations thereafter, as well as from the company's diversified assortment of products. To help solve these developments, a varied "assortment" of internal and outside specialists was added to the networks on the way. All this made interaction structures more difficult and coordination issues more important.

By delegating co-responsibility to the business unit level, decision making became more "vertical" and added an additional management level made up of various vertical and horizontal sub-clusters, and consisting of permanent and temporary cooperative organizational structures. The business unit, made up of its product management groups and specialists, was one of them. Additionally, various joint structures were active, such as NSC's Management Committee, the Marketing Committee (made up of all business unit managers and the managers of the B brand units), the Advertising Committee, and various national and international intra-Philips working groups, in which the Research Group, the Production Groups and Multinational Staff Specialists were also represented. Therefore, decision making developed into very time consuming consultation rituals. This also resulted into the unwanted side effect that individual managers became less willing to accept individual responsibility. However much – as from 1978 – NSC's newly entered CEO tried to turn business unit managers and their deputies into entrepreneurs, that learning process did not come about, because the managers generally felt that the new CEO was predominantly putting on a good internal testimony show.

Consultation became a major official ideology based upon harmony and consensus. To a great extent it was a myth. Most managers did interact in networks, indeed, but their roles and interests often were so much apart that it was just too normal for their views, perceptions and intentions to be different from and to conflict with one another. However much disapproved and unwanted officially, bargaining and negotiating became the unofficial ideology. This was legitimized in daily practice, as long as it was ritually called "consulting."

Whenever self-interest made it necessary, the NSC also had her fair share of diplomatically disguising things to say to the outside world. The company normally behaved "manly," she was always said to know what she was talking about, and to be in full command of the situation. Outsiders who were not adequately familiar with the inside situation often found it hard to decode statements made by NSC's management. In many cases statements were phrased in such a way that they could mean different things. Sometimes management seemed to "take the bull [the issue concerned] by the horns" by formulating questions without giving the answer,[2] or responded by raising counter-questions.

It can be said that, in summary, during the sixties to the early eighties some major trends are noteworthy regarding norms on perception and action, how managers were supposed to perceive strengths and weaknesses, opportunities and threats, how they should deal with them (in cooperation with or in opposition to inside and outside actors), and how they should respond to them.

It was a sellers' market in 1965; Philips products found easy buyers. Managers did not (have to) think about change, new and weak environmental signals and business ethics. For these self-confident managers sales and profits were just the natural thing of the day. Philips decided what to make and sell. Decisions were taken in a rather authoritarian way. By education and experience, the NSC was focused on markets only.

Since then with student revolts, oil crises, consumerism, environmentalism, Japanese competition, recession and more government rules and sales and profits feeling the draught, management staff levels developed a more critical, ethical and societal awareness. A concensus ideology established itself officially, and a negotiation ideology was the informal order of the day. Management relationships became asymmetrically "open," i.e., some managers monopolized (some) relevant information. Conflicts were called harmony, and compromise was labeled concensus.

Internally, as well as externally, the ideology resulted in responsivenesses that were mainly dynamically conservative or testimonial. The language used suggested otherwise, and presented the company as analytically and implementingly responsive. In niches of business policy and in a limited way during the seventies and early eighties, the NSC and the business unit learned to actually become (more) analytically and implementingly responsive.

Around 1980, a new business sense and pragmatism were born. Managers were (increasingly willing and) learning how to manage large scale operations in a more flexible and small scale way, and also became more societally intelligent. They started to face the outside world more actively and openly. This was the beginning of organizational learning, and of business managers starting to learn (how) to integrate new developments into their own policies, instead of making themselves "immune," for their own weaknesses and for outside uncertainties, by delegating this part of their job to specialized others.

Table 5.1 summarizes values and beliefs, attitudes and perceptions of social realities of the NSC and the business unit, as they changed in the sixties to early eighties.

Political Sub-relationships

Thanks to education, experience, in-company training courses, as well as active networking inside the business unit, NSC, the rest of Philips and the outside world, managers became more professional in the seventies. In spite of that, however, they were not able to prevent and solve situations of zero and minus growth. And in

TABLE 5.1. Values and Beliefs, Attitudes and Perceptions of Social Realities of the National Sales Company (NSC)

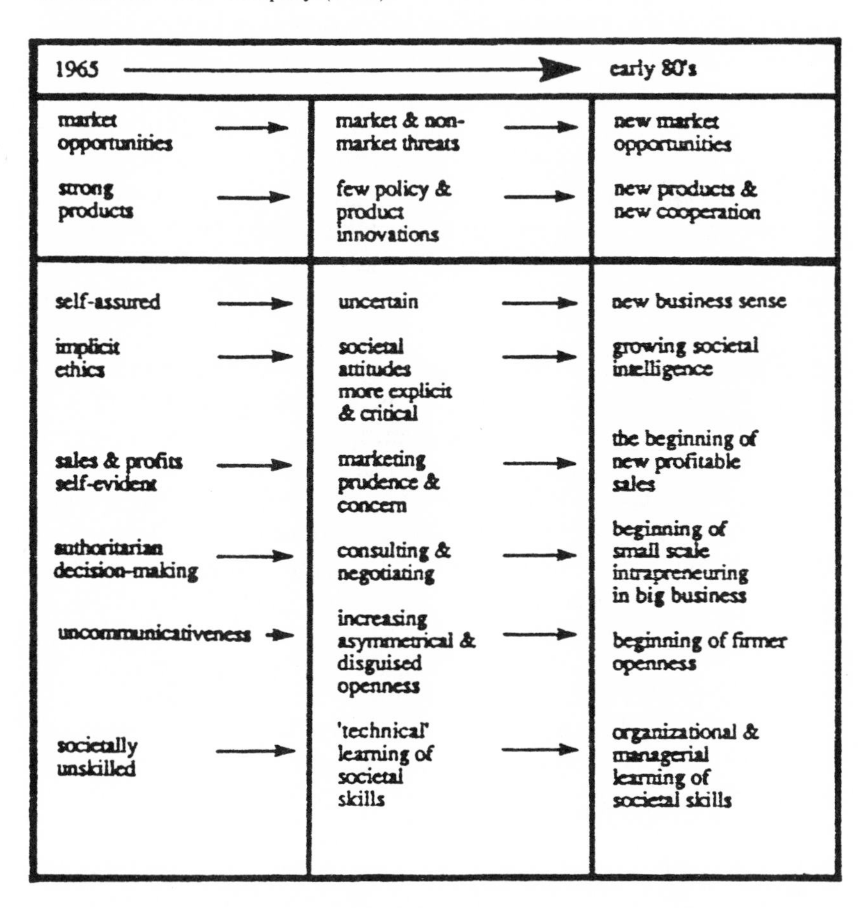

1965 →		early 80's
market opportunities →	market & non-market threats →	new market opportunities
strong products →	few policy & product innovations →	new products & new cooperation
self-assured →	uncertain →	new business sense
implicit ethics →	societal attitudes more explicit & critical →	growing societal intelligence
sales & profits self-evident →	marketing prudence & concern →	the beginning of new profitable sales
authoritarian decision-making →	consulting & negotiating →	beginning of small scale intrapreneuring in big business
uncommunicativeness →	increasing asymmetrical & disguised openness →	beginning of firmer openness
societally unskilled →	'technical' learning of societal skills →	organizational & managerial learning of societal skills

new (consumerist) situations managers were not willing and able to initiate and implement new policy responses successfully. For example, management's insights fell short when it came to managing energy, noise and product life challenges and turning them into company strengths. Their power and influence were limited in those respects.

Not everything managers were doing was well-considered and consistent. For example, as soon as a Production Group finalized a new product, they felt they could not wait to market it. At times the

NSC or the business unit did not believe in a new product, but most of the time they could not stop its market introduction for long. Voluntarily or from being compelled, it was felt to be in the longer run internal political interest of the business unit to use all the business power it could to help mobilize market demand. If, on the other hand, consumerists were asking the business unit to develop and implement new societal responses that would also include an effort of mobilization of consumer behavior, then the very same business unit was strong enough not to take notice of it, arguing that its power was insufficient for such a mobilization task. It normally was much easier to say "no" to external actors than to internal ones.

Formally speaking, the business unit's management decided about the unit's policies. Real decision making, however, was largely a matter of "social engineering" in networks. Management's responses mostly were opportunistic "small steps" only. They largely resulted from perceived external and internal power relationships. For example, the annual budgetary sessions up and down the line were ritual political "dancing" to a great extent, borrowing power from others, exchanging gains and sacrifices. From 1965 to 1975, NSC's management was still in a position to dictate policies, or simply approve or disapprove plans presented to her. Since 1975 – a time by which business unit management was firmly established – unit management no longer accepted unfounded conclusions by NSC's management. Respectively, the business unit's management only implemented unacceptable strategies and policies conditionally. If it thought that NSC's management was setting goals that were too high then it would ask for more financial means to "beef up" the activities to realize those goals. In case NSC's management wanted unit costs to go down unrealistically, then unit management did not accept her full co-responsibility for the targeted results any longer.

Managers knew that their careers strongly depended upon judgement by peers and superiors. In this respect the political climate developed in such a way that it became increasingly difficult to openly call a mediocre manager mediocre, knowing that at some (unmanageable) time and place, the boomerang might come back and hit oneself (even harder). Training courses had her share in the judging processes. For a long time the training system formed one-sided managers, fit for growth situations, not aimed at educating creative entrepreneurs capable of responding to new and more complicated types of change.

In 1979, NSC's management proved strong enough to prevent the undermining of its traditional, internal power base. The integration of (marketing) quality control into the company's policies and activities was prevented (Divita, 1978, pp. 74-78). Management also proved strong as compared to the consumer staff specialist of the multinational's Chief Executive Board, by keeping away from the NSC an initiative aimed at starting systematic management of product complaints at the NSC level, in order to improve product policy.

As said in the "Introduction" of this section, the Production Groups were responsible for developing new products for all of the international markets. In case new products would have to be specially tuned to the wishes of the Dutch market, that would mostly cost at least 10% more. Together with France, West Germany and the U.K., Holland generally speaking, was one of Philips' "key countries" in Western Europe. This was so because of sales volume, market share, profitability, or for external or internal trendsetting reasons.

In terms of products, the picture for Philips in the seventies in Holland was a varied one, ranging from the best in the market, via "me too," to dated.

Price was being used by the business unit to cash in on unique products or to compensate for minor products, or to offer better margins to the trade than competitors were doing, particularly when competitors offered sub-standard margins. Vis-à-vis the trade generally, the business unit was practicing a differentiated and complex pricing policy in such a way that it was hard for others to pinpoint what Philips was actually doing. Quite often these practices were seen as "divide-and-rule."

Philips consumer advertising in Holland proved to be quite effective, particularly so because of her dominant advertising budget and its wide presence all over the country for so many years. Qualitatively, the Philips' advertising impact suffered a great deal from being unpersonal and faceless, and not interesting and adventurous enough for the younger generation. Inside Philips, dealer awareness, retailer oriented responsiveness and (more) sophisticated merchandising were just about to begin to take shape.

Philips product warrant conditions and after-sales servicing got good marks generally. The company's wide presence and dominant market shares in the country also served the purpose of confirming and strengthening the strong Philips brand further still.

By 1969 consumers were still prepared to accept much from manufacturers and retailers. By 1979 the very same consumers had learned (how) to complain: sooner, more, stronger and—if necessary—more lastingly. The NSC had proved strong enough to turn a relative weakness into an additional strength, by starting a number of consumer departments in the seventies, for handling consumer complaints and for giving individual information. Business unit management did not, however, find herself strong enough to make consumers more societally responsible, neither under her own (Philips) wing, nor jointly with other firms in the business, nor with the government and the consumer unions. For example, however good it would be to develop a lower wattage vacuum cleaner, as long as consumers take higher wattage to stand for a better product quality—resulting from the long time marketing efforts of all vacuum cleaner manufacturers combined—it would be no good to do so. There were only a few strong instances that made the business unit management worry, notably new government rules and regulations and publicity of strongly negative results of comparative product tests of the consumer unions because first, there was a tendency of retail management to tune their purchases more to the positive and negative results of these tests, and secondly, because these test results often proved to influence new product introductions considerably. As to Philips activities generally, the consumer unions had a very limited impact only. In fact the consumer unions became part of the establishment themselves. The NSC and the business unit were playing it societally intelligent therefore by stating time and again that Philips "had learned to listen well to the Consumer Union."

By the late seventies the NSC understood the importance of maintaining good relations with all relevant outside actors, market and non-market. As to government circles—including advisory councils – the NSC management started to build high level contacts in a more systematic way. By doing so, a useful platform was realized for "times when governments do not give an issue sufficient consideration." Thanks to that platform, the NSC could more easily and comfortably explain to the government, that even the "Great" Philips is not always strong enough on her own to integrate new societal issues in her policies and activities without negatively affecting its competitive power and position. The platform sometimes also nicely worked the other way round. Following the second oil crisis of 1978/79, the NSC asked the Netherlands Minister of Economic Affairs which steps the Administration had in mind to secure electricity supplies. Respondingly, Philips was asked by the Minister in return to publicly give herself a "higher energy profile," which the company found impossible (or unpractical) to refuse. The booklet "Philips and Energy," that resulted from this informal governmental move implied to the reader that it was a Philips initiative.

One may first conclude that the relative power of NSC's business units' managements and individual managers were situationally set. As compared to other Philips managers inside and outside of the NSC, and to outside actors, managers were strong in some issues sometimes, but not in all issues all of the time. Individual managers may make and take decisions individually in the formal sense of the word, in fact this mostly meant that they were formalizing decisions that had already been "engineered" and agreed upon before jointly. Sometimes managers succeeded to push their own opinions and response intentions through in spite of (much) opposition, but most of the time decisions made were compromises. This was because however powerful a manager may be in principle to initiate new response intentions and proposals, he cannot mostly force others to accept and implement those responses if the others do not agree and are not really willing to cooperate, whatever they seem to be.

Second, business unit management no longer accepted everything from NSC's management. When NSC's management wanted to push something through, the business unit often made many provisos; whenever possible the unit implemented the policies in her own self-fulfilling ways. As a result, NSC's management found herself stuck with the black Jack. In such situations, chances for the business unit to get her policy recommendations approved the next time largely improved. As NSC's management became aware of this, it often only "polished" the policy recommendations it received, or informally assured beforehand that no policy plans would be submitted that would be out of the way from accepted thinking. Externally, similar developments were observed. Individual consumers did not simply accept every (new) product offered anymore and the members of the Consumer Union did not act in the normative rational ways consumerists were supposed to.

Third, internally it was not so simple anymore to use one's power in a direct way. Managers were increasingly looking for new ways to use their power in more informal and indirect ways, e.g., by selectively channeling relevant information, rewarding business unit managers, and (subtly) using career dependencies. Externally, indirect methods had been used much longer. Consumers have tried to influence producers indirectly for quite some time via the consumer unions, governmental authorities, and the media. The other way round, the NSC sometimes tried to influence consumers via the editorial columns of consumerist periodicals. Philips intensified its political contacts with the same thing in mind.

Fourth, by the late seventies, external contacts of NSC's management strongly aimed at legitimating the NSC, trying to prevent the government and consumerists from intervening in the business practices of the company in order to stabilize and support the company's power positions. The more it succeeded in this respect, the more the business unit could limit herself to smaller step responses only. So our fourth summarizing conclusion is that NSC's culture and power relationships particularly tried to secure continuity of its policies. Within this framework business unit managers had a great degree of freedom, as long as they remained within the boundaries of established ways of thinking and doing, and as long as they complied with accepted norms. Only unique personalities were exceptions to the rule. So it may not be fair to make business unit managers responsible for integrating consumerist developments into their business policies, when at the same time curtailing their power to do so. It is not before the early eighties that one of NSC's top managers had come to the conclusion that this managerial "blocking" had a negative effect upon the unit's responsiveness. As from that moment, the NSC began to formulate policy intentions about the kind of information needed and the way in which information processing should be adapted and organized.

Fifth, managers of the business unit and of NSC's management team mostly worked informally and speedily. Whenever decision making processes involved interactions between NSC units, with other operating companies, and with the outside world at large, often it was difficult and time consuming for the business unit to have it its way. Particularly so if it had not learned to manage short-cut, informal networks.

Sixth, however powerful management might be, actual influence greatly depended upon the willingness to use that power. When self-interests were well served by using power, then the propensity to use it in culturally permitted ways—including creative efforts to strengthen the power base as such—was far greater generally. As far as managers did not perceive that self-interest or were not forced to see it, at best managers tried to legitimate their conservative responses, "acting" to be unable to do otherwise. Most of the time, the NSC and the business unit were practicing responsivenesses of a testimonal nature. Sometimes managers followed their more societally engaged personal feelings and formulated more responsive alternative policy intentions. Quite often these more responsive alternatives, however, were presented to NSC's management in a far less convinced and convincing way than "traditional" proposals. So the more responsive solutions did not really get a fair chance to be accepted most of the time. Possibly those outcomes were very much in accordance with the deeper feelings (and fears) of the initiators anyway.

CONCLUSIONS

Finishing this essay, let us conclude what we have learned and do so in a decreasing order of generality.

First, our participating observations inside Philips—based upon our dynamic theoretical model—made us understand the learning and decision making processes under review. We learned that managements and managers have learned or were beginning to learn how to deal with unfamiliar types of change and that they were doing so in ways similar to our model, mostly unaware of the processes they were going through on the way. When facing them with our model for the first time, managers often did not recognize their own behavior, or even denied the relevance of parts of it. Then it was surprising to find how they experienced sort of a shock of recognition when having a second look at the model. In fact, what the model does is to face a manager deeply with his multifaceted, purposeful complexity of integrating people in action. The model was also valued as a useful tool to create a knowledgeable mind set and train the wit of managers quicker and at less expense. By applying the model intelligently and creatively, it is easier to understand one's own historically and situationally grown organizational settings, culturally and politically and to make them work for one's self better. Through the model, managers can be faced with their willingness and ability to keep on learning new "know-whats," "know-whys," "know-hows," and to teach others correspondingly, to really risk one's neck and do it, or not do it.

Apart from an exceptional "response champion" at middle or lower management levels here or there, business units, divisions and companies will only aim at and materialize a right kind of response processes and types of responsivenesses, when top-managements are convincingly inciting and managing them. It can only happen if top-managements are strong educators of change; if the strategic changes chosen are (made) feasible; if top-management is setting the example and is allocating enough time, money and people; and if people and systems are changed, amongst others through retraining, new reward systems and (cross-functional and international) job rotation schemes. Organizational culture can only be changed after changing top-management's culture.

Second, culture at the top- and upper-management levels developed along identical lines; not only so because all of them dealt with identical spirits of the time and facts of life, but also because they have taught each other and have learned from each other in terms of role values and beliefs, norms and behavior. Technical product input orientations continued to be dominant in the company everywhere. In the Netherlands Sales Company this orientation was overshadowed by a sales output focus, in which competition and the trade often played a greater role than the consumer. The "me" oriented culture was stronger than the "you-and-me" one most of the time.

In the sixties and seventies science in Philips was not as successful as expected. That triggered a feeling of uncertainty, of a problem trauma even: "Problems, problems, problems, we've got enough of them, please no more." It also lead to anti-intellectual reactions of increasing interest for "softer" type approaches through feeling, seeing, and understanding things. This was not the type of climate that stimulated "response champions." "Fail safe" was the order of the day, not "safe fail."

This was also not the time for high profile leaders. Leaders should be – at least should look like – non-leaders. They should imply freedom of choice and action for others, harmony and concensus. By behaving that way the company ran the risk of more uncertainty and of conflict having their way, unless prevented or cured in informal and invisible "social engineering" ways. Openly using power – if that were not a contradictory thing by itself – was very much associated with misusing power, and therefore a cultural taboo.

Managers increasingly understood that they could only have it much their way, when packing their thinking and doing in a "you-friendlier" way: in ways sympathetic to the boss, peers, lower managers, consumerists, government authorities, and showing great talents in adopting jargon coined by others. The Netherlands Sales Company may have done away with a great deal of formal and written procedures, may have delegated part of top-management's job to a newly set up upper-management level or to staff specialists at the top or upper level, that did not mean that bureaucracy and hierarchy had become looser. In fact it resulted in a blending of less bureaucracy with regard to smaller and more familiar types of decisions, more bureaucracy regarding very major and unfamiliar responses, and an unbureaucratically packed bureaucracy at large.

Third, in terms of responsiveness and responsive roles, the business unit management learned very little from the Chief Executive Board or the CEO of the multinational Group of Philips companies until the early eighties because the CEO and his Board virtually abstained from missionary response activities inside the corporation. If not invisible, their in-company communications were unclear, unconvinced and uncertain. In those respects leadership also followed a low profile policy. They were not (yet) aware of the important role they had to play as cultural educators. This low profile attitude partly followed from the prevailing philosophy of the times, according to which Philips was considered to be a federation of companies, operating highly autonomously. In some ways top-management of the Netherlands Sales Company prevented important consumerist information—sent by Board level specialists—from reaching the business unit, and behaved like a blocking and filtering gatekeeper. And last but not least, there was hardly any Board oriented social intelligence at the business unit level. "The skies were high and the Emperor was far away."

The business unit management did, however, constantly learn from top-management of the Netherlands Sales Company, as well as from its peers, staff and specialist advisers, who in a way formed sort of an abstract joint teaching leadership. These references made business unit management wonder whether it was functioning the way it should and whether its policy should follow a culturally and politically comfortable line of "more-of-the-same" or of small step changes. Bigger step change proposals were quite incidental ones, because top-management of the Netherlands Sales Company not only blocked consumerist information, but also filtered information about important political developments in the country. Business unit management never got "big step" change proposals accepted. In that respect the two management levels were incompatible. With regard to responsiveness and responses, top-management was split into three factions. The member of the top-management team, to whom the business unit management was directly responsible, represented all of the characteristics of conservative responsiveness. Testimony, analysis and implementation responsiveness were looked upon favorably only as long as they would help "Sales" to do their proper job. Another member of the top-management team was responsible for all of the marketing staff services of the operating company, including the handling of consumer complaints, consumer information, and the high level contacts with governmental authorities and consumerist organizations. This top-manager's societal intelligence was much more developed, and so was his in-company role. But as he had only co-responsibility for sales, the business unit's social intelligence was not primarily aimed at him. Third but not least, the Chairman of the top-management team very much played a refereeing role closer to Sales than to Staff. Business unit management learned that one should carefully handle differences of opinion, and that it was a cultural must to avoid direct confrontations.

In 1982, when a new CEO stepped in at Headquarters, an interesting entrepreneurial era headed by a charismatic leader began. But that is another story.

NOTES

1. During the closed meeting with ambassadors on January 22, 1981, and inspired by Walters (1975) and Weinstein (1976), Dekker—member of Philips' Chief Executive Board—said amongst others:

> The businessman's responsibility is to keep his business going and to guarantee continuity . . . It becomes increasingly difficult to separate business from politics. The enterprise cannot avoid—at least—being conscious of politics and forming an opinion about them . . . If it is true that we shall slip from a mixed economy towards a form of state socialism, if managers do nothing (underlining by Dekker), then in order to reverse that development we shall have to change the attitude and action of managers so as to influence (underlining by Dekker) coming events. This implies that something must be done about the knowledge of and experience in and with politics . . . The self-image, as non-political, which businessmen adopt, is derived from nineteenth century ideas which have become anachronistic. The businessman must anticipate developments and that means acquiring an understanding of the viewpoints of others. Political analysis should be a standard part of the company's strategy, the aim being to present positive alternatives to other groups and to influence the climate within which the politics are determined . . . Businessmen's work is political in the sense that (1) it depends on and gives support to the distribution of authority, power and remuneration, (2) businessmen allocate and re-allocate the resources available to them in such a way that they have an important influence on the lives of those working for or with them, and (3) businessmen are continually faced with finding solutions to conflict situations.

2. One of the members of NSC's top management team (Kuijpers, 1979, pp. 21 and 25) publicly formulated the following questions:

Which are the new product development requirements of the outside world? Trade unions are interested in employment and working conditions. The central government and local governmental authorities are concerned about environmental effects of manufacturing processes. And how does the situation look like with regard to raw materials and energy? Is the product a safe one? Philips has got the know-how to answer many of these questions. On the other hand, are we really able to pinpoint and formulate in a nuanced and quantified way which consumer behavior will be relevant for new product development and which will be the market needs and wants of the future?

REFERENCES

Ackerman, Robert W. (1973). How Companies Respond to Social Demands. *Harvard Business Review*, July-August.

Ackerman, Robert W. (1975). *The Social Challenge to Business*. Cambridge, MA: Harvard University Press.

Ackerman, Robert W. & Bayer, Raymond A. (1976). *Corporate Social Responsiveness – The Modern Dilemma*. Reston VA: Reston Publishing.

Ackhoff, R.L. (1974). *Redesigning the Future – A Systems Approach to Societal Problems*, NY: Wiley.

Advertising Age (1974). Special Section, Energy Crisis, February 4.

Ahold CEO: Reply letter of June 5, 1980 of Holland's leading food store in which our invitation to participate in longitudinal in-company research was declined.

Annual Reports of Philips 1968, 1973, 1974, 1976, 1978 and 1979.

Ansoff, H.I. (1974). The Future of Corporate Structure. Working paper No. 74-4, abridged version, Brussels: European Institute of Advanced Studies in Management EIASM.

Ansoff, H.I. (1975). Managing Surprise and Discontinuity – Strategic Response to Weak Signals. Working paper No. 75-21, Brussels: European Institute of Advanced Studies in Management EIASM.

Ansoff, H.I., Declerck, R.P. & Hayes, R.L. (1974). From Strategic Planning to Strategic Management. Working Paper No. 74-32, Brussels: European Institute of Advanced Studies in Management EIASM.

Bauer, Raymond A. (1974). An Agenda for Research and Development on Corporate Responsiveness. Graduate School of Business Administration, Boston, MA: Harvard University (unpublished).

Brand, F.L. van den (1980). Philips Human Resources Manager in interview in leading Dutch newspaper NRC Handelsblad, Rotterdam, February 9.

Brodèn, P. (1976). Turbulence and Organizational Change. Linköping University, Sweden.

Burns, T. & Stalker, G. (1961). *The Management of Innovation*. London: Tavistock.

Chandler, A.D. (1962). *Strategy and Structure*, Cambridge, MA: MIT Press.

Child, J. (1972). Organizational Structure, Environment and Performance—The Role of Strategic Choice. *Sociology*, January.

Child, J. (1973). Organization—A Choice for Man. In Child, J. (ed.), *Man and Organization*.

Cyert, R.M. & March, J.G. (1963). *A Behavioral Theory of the Firm*. NY: Prentice-Hall.

Dekker, W. (1978). Member of Philips' Chief Executive Board, lecture, Diversified Applications of Electronic Technology and Management, International Seminar on Prospects for the Electronics Industry for the Next Decade in the Light of Technological Innovation and Philosophy and Corporate Management, Tokyo, October 12-13.

Dekker, W. (1979). Member of Philips' Chief Executive Board, lecture at the annual meeting of Philips' Young Graduates Association, Will European Enterprise Make It To 1985? Eindhoven, November 29.

Dekker, W. (1981). Member of Philips' Chief Executive Board, lecture at closed meeting with ambassadors, Holland, January 22.

Dekker, W. (1982). Philips CEO, interview with the newly appointed CEO, Philips Koerier, January 14.

Dierkes, Meinolf & Bauer, Raymond A. (1973). *Corporate Social Accounting*. NY: Praeger.

Dijck, J.J.J. van (1979). Structuur, besluitvorming en conflict—een organisatiesociologische verkenning, in: J. Heijnsdijk en Meyaard, L.J. (eds.), Organisaties . . . een toekomstperspectief, Samson, Alphen a/d Rijn (Holland).

Dijck, J.J.J. van (1981). Beheersing van sociale ontwikkelingen, in: Dijck, J.J.J. van (ed.), Beheersing van sociale ontwikkelingen—naar een sociaal ondernemingsbeleid, Stenfert Kroese, Leiden (Holland).

Divita, Salvatore F. (1978). Marketing Quality Control—An Alternative to Consumer Affairs. *California Management Review*, Vol. XX, no. 4, Summer.

Dye, T. (1972). Understanding Public Policy, NY: Prentice-Hall.

Easton, D. (1965). *A Framework for Political Analysis*, NY: Prentice-Hall.

Emery, F.E. & Trist, E.L. (1965). The Causal Texture of Organizational Environments. *Human Relations*, Vol. 18, February.

Emery, F.E. & Trist, E.L. (1972). Towards a Social Ecology, London: Plenum Press.

Frankfurter Allgemeine Zeitung (1973). Article on energy by H.W. Bodmann, December 28.

Hakansson, H. (1982). International Marketing and Purchasing of Industrial Goods. Chichester; Wiley.

Hamersveld, M. van (1977). Staff Adviser of Philips' Chief Executive Board, lecture at Eindhoven University of Technology, Customer Slave or King?, Eindhoven.

Jenkins, W.I. (1978). *Policy Analysis—A Political and Organizational Perceptive*, London: Robertson.

Klooster, W. van der (year of publication unknown). The Environment and the

Marketing System. Marketing Training Group of Philips' Chief Executive Board, Eindhoven.

Klugt, C.J. van der (1980). Member of Philips' Chief Executive Board, in-company lecture on New Product Development, Eindhoven, February 14.

Kotler, Philip (1973). The Major Tasks of Marketing Management. *Journal of Marketing*, October.

Kotler, Philip (1988). *Marketing Management – Analysis, Planning and Control*. Englewood Cliffs, NJ: Prentice-Hall, first edition 1967, third edition 1976, fourth edition 1980, fifth edition 1984 and sixth edition 1988.

Kuilman, M. (1980). Member of Philips' Chief Executive Board, interview in leading Dutch newspaper NRC Handelsblad, March 8.

Kuijpers, A.C.H. (1979). Member of the top-management team of Philips' Netherlands Sales Company, Produktontwikkeling en onderzoek naar consumentengedrag, Tijdschrift voor Marketing, March.

Lawrence, P.R. & Lorsch, J.W. (1967). *Organization and Environment – Managing Differentiation and Integration*. Cambridge, MA: Harvard University.

Lawrence, P.R. & Lorsch, J.W. (1967). Differentiation and Integration in Complex Organizations. *Administrative Science Quarterly*, Vol. 12, No. 1.

Loupart, O.M.E. (1958). Member of Philips' Board of Directors, Memorial Book of Philips' Young Graduates Association, lets over Philips, Eindhoven.

March, J.G. & Simon, H.A. (1958). *Organizations*. NY: Wiley.

Mill, C. Wright (1972). Language, Logic and Culture. In: Cashdan and Crugeon (eds.), *Language in Education – A Source Book*, London, Routledge and Kegan Paul.

Murray, E.A. Jr. (1974). *The Implementation of Social Policies in Commercial Banks*. Cambridge, MA: Harvard University.

Murray, E.A. Jr. (1976). The Social Response Process in Commercial Banks – An Empirical Investigation, *Academy of Management Review*, July.

Netherlands Employers Federation (1978). Commissie Consumenten Vraagstukken VNO/NCW, Minutes of internal meeting dealing with how to reply to our Project Proposal on Corporate Responsiveness, The Hague.

Noordhof, D. (1970). Member of Philips' Chief Executive Board, lecture at the 187th Annual Meeting of the Netherlands Society for Industry and Trade, Sociale problemen die voortvloeien uit de veranderingen, die zich in en om de onderneming voordoen, Utrecht (Holland), June 11.

Otten, F.F. (1980). Member of Philips' Chief Executive Board, interview in leading Dutch newspaper NRC Handelsblad about major reorganization plans, March 8.

Otten, F.F. (1981). Member of Philips' Chief Executive Board, lecture, Dialoog met de samenleving en marktbeînvloeding – een spanningsveld? Annual Conference of the Netherlands Advertising Society, Utrecht (Holland), March 20.

Pettigrew, A.M. (1973). *The Politics of Organizational Decision-Making*, London: Tavistock.

Philips, Frits (1976). 45 jaar met Philips, Ad Donker, Rotterdam.

Philips' General Advertising Department (1976). 22 Questions—A Touchstone for Advertising Standards, Eindhoven, February.

Philips' Production Group Lighting (1974). *Lighting News*.

Philips Nederland (Netherlands Sales Company (1980). Publication on energy "Philips en energie." Eindhoven, August.

Rees, J. van (year of publication unknown). The Marketing Concept with Philips, Marketing Training Group of Philips' Chief Executive Board, Eindhoven.

Rhenman, E. (1973). *Organizational Theory for Long-Range Planning*, London: Wiley.

Riemsdijk, H.A.C. van, Jhr. (1973). Philips' CEO, speech at Annual Meeting of Shareholders, Aspecten van het ondernemerschap, Eindhoven, April 18.

Riemsdijk, H.A.C. van, Jhr. (1974). Philips' CEO, speech at Annual Meeting of Shareholders, Philips in de ontwikkelingslanden, Eindhoven, April 18.

Riemsdijk, H.A.C. van, Jhr. (1974). Philips' CEO, speech at a meeting of Philips' Internal Consultation Body in the Netherlands, Philips as an international enterprise, Eindhoven, April 17.

Riemsdijk, H.A.C. van, Jhr. (1975). Philips' CEO, speech at Annual Meeting of Shareholders, Onze onderneming onderweg, Eindhoven, April 24.

Rodenburg, N. (1977). Philips' CEO, lecture for the Netherlands Society for Industry and Trade, Philips in de wereld, Groningen (Holland), November 22.

Rodenburg, N. (1978). Philips' CEO, speech at Annual Meeting of Shareholders, Op zoek naar nieuwe evenwichten, Eindhoven, May 9.

Rodenburg, N. (1978). Philips' CEO, lecture at 1978 Philips Young Graduates Association, Internationale politiek, economie en ondernemingen, Eindhoven, December 12.

Rodenburg, N. (1979). Philips' CEO, speech at Annual Meeting of Shareholders, Philips in het licht van de gloeilamp, Eindhoven, May 8.

Schreyögg, G. (1978). Kontextbedingungen als Determinanten der Organisationsstruktur, Erlangen-Nürnberg.

Silverman, D. (1968). From Formal Organizations of Industrial Sociology—Toward A Social Action Analysis of Organizations. *Sociology*, Vol. 2.

Silverman, D. (1970). *The Theory of Organizations*. London: Heinemann.

Simon, H.A. (1957). *Administrative Behavior*. MacMillan.

Starbuck, W. (1975). Organizations and Their Environments. In M. Dunnette (ed.), *Handbook of Industrial and Organizational Psychology*, Chicago.

Terreberry, Shirley (1968). The Evolution of Organizational Environments. *Administrative Science Quarterly*, Vol. 12, No. 4.

Thompson, J.D. (1967). *Organizations in Action*. NY: McGraw-Hill.

Turnbull, P.W. & Cunningham, M.T. (1981). *International Marketing and Purchasing, A Survey Among Marketing and Purchasing Executives in 5 European Countries*, London: MacMillan.

Tuyt, J.J. (1981). Chairman of the top-management team of the Netherlands Sales Company, lecture, Consumentengedrag en ondernemingsstrategie, Amsterdam, November 26.

Chapter 6

Sociopolitical Analysis for International Marketing: An Examination Using the Case of Egypt's "Infitah"

Essam Mahmoud
Gillian Rice

INTRODUCTION

The marketing of products and services internationally is conditioned by environmental factors in addition to being dependent on consumer characteristics. The focus of this chapter is on the sociopolitical environment in Egypt. The sociopolitical environment is the noneconomic environment which encompasses those elements of human society that manifest themselves in political change. Political change is important for the international marketer when it results in the implementation of policies that affect trade and investment. See Figure 6.1 for a schematic representation of this process.

This chapter is organized as follows. First is a brief examination of the nature of the sociopolitical environment as it pertains to international marketing. Second is a review of some frameworks for the analysis of the relationship between sociopolitical factors and business. Next, these frameworks are evaluated in the light of their ability to describe the experience in one country, Egypt. Finally, suggestions for further developments in sociopolitical analysis are made.

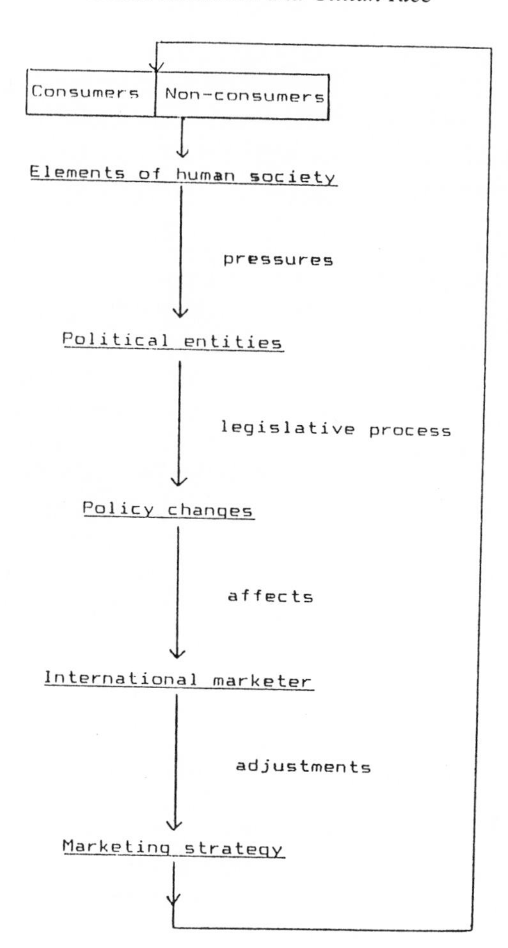

FIGURE 6.1. Sociopolitical Processes and the Implications for International Marketers

THE SOCIOPOLITICAL ENVIRONMENT AND INTERNATIONAL MARKETING

The sociopolitical environment and its implications for business have most often been analyzed at the "macro" level. This assumes that sociopolitical change affects all business in the same way. Of course, this is not necessarily true. Robock (1971) was the first to distinguish between macro and micro risks. Environmental risks at the micro level are those relevant to a particular firm or project only. Brewer (1985) argued for a multilevel analysis of the interactions of political and business systems. For example, he identified four levels or situations:

1. a macro level political event, such as a change in the party in power, can have macro level risk consequences for all firms if the new government wants to renegotiate all tax holidays for all activities of foreign firms;
2. a micro level political event, such as a government official changing his attitude about trade sanctions, could create micro risks for particular firms' exports of a particular product;
3. a macropolitical event, such as an international conflict, can create micro risks for particular projects if imported inputs for the project are subject to selective trade sanctions;
4. a micropolitical event, such as a particular minister increasing his bureaucratic influence, can create macro level risks if he is able to implement new tax policies affecting all firms.

Thus, the international marketer must be concerned about the impact of environmental occurrences on his firm's operations, whether these occurrences are the result of a violent change in political regime or a gradual process of social and political evolutions (de la Torre and Neckar, 1988). If the international marketer would like to anticipate possible policy changes of relevance to marketing strategies, he must understand the underlying sociopolitical processes. Nonpolitical trends (such as social and economic trends) and the political response are interdependent (Ascher, 1982). Governmental policies are the political outcomes conditioned by socioeconomic trends. Ascher suggests that policy forecasting is contingent upon identifying social trends which:

1. change the priority of issues facing the political body;
2. change the resource bases of different political factors (such as increased wealth or prestige which are convertible into political leverage);
3. activate different political dynamics than those currently prevailing.

Therefore, as emphasized by Austin and Yoffie (1984), political events should not be evaluated as independent acts, but as part of a country's historical situation. By analyzing the structure of a society and its government, political coalitions and bureaucratic inertia, a manager is better equipped to judge the probability and direction of relevant policy shifts. In this context, various approaches to sociopolitical analysis for international business have been suggested by researchers and practitioners.

ANALYTICAL FRAMEWORKS OF SOCIOPOLITICAL FACTORS AND THEIR IMPACT ON BUSINESS

Three analytical frameworks are discussed. These are chosen for their diversity and are: Gillespie's coalition cycle, bargaining power models and de la Torre and Neckar's framework for political risk forecasting.

Gillespie's Coalition Cycle

Gillespie's (1984) approach was developed to explain Egyptian foreign economic policy in the Sadat era. She uses a conceptual framework of the relationships between three interest groups: foreign investors, government and local investors. Gillespie identifies a "coalition cycle" as various bipartite coalitions among these three interest groups that form and dissolve over time. The changing power of some interest groups relative to others in a country can affect the permitted ways of doing business.

The coalition cycle can be summarized as follows (Gillespie, 1984). The "development coalition" occurs when government and foreign investors form a bipartite coalition. The government grants foreign investors privileges beyond those given to local private investors in return for certain advantages that the foreign investors offer. These advantages include capital, technology, foreign exchange and market access. The government's rationale arises from a need to satisfy local interest groups other than local private investors. The inevitable conflict in goals, however, (for example, the foreign investor seeks profit; the government seeks national economic development) strains the coalition and allows the excluded third party (the local private investors) to take advantage of the rift between the original two parties. The development coalition becomes a "capitalist coalition" as foreign and local investors form joint ventures. Foreign investors may need local market knowledge such as distribution systems. Local investors may need access to technology, trademarks and a potential ally against local or international competition. As the government may not be satisfied with the response by foreign investors to its economic liberalization policies, it may allow concessions to local private investors. Sometimes one outcome of the capitalist coalition is the attempt to influence the government to encourage private enterprise in general.

As local private investors recognize a strengthening of their position, they will begin to lobby for a "nationalist coalition." Changes in the investment environment include more local ownership and especially in the case of foreign-local joint ventures, the local position will be increased. Local investors, however, may avoid capital intensive projects or projects that require a number of years to produce profits, despite the fact that the government wants such projects in order to develop the country. Local investors tend to take advantage of the higher margins to be made by selling luxury goods to the rich, for example. Therefore, other interest groups in the country will begin to oppose the nationalist coalition and may call for a new development coalition. This would result in the value of an investment to the nation as a whole being more carefully scrutinized. The coalition cycle may begin again. One coalition cycle in Egypt was identified by Gillespie (1984) during Sadat's period.

Bargaining Power Analysis

Bargaining power models use social cost-benefit analysis to identify appropriate investment strategies, usually in developing countries. According to Vernon (1971, 1977), for example, the relationship between host developing countries and multinational enterprises (MNEs) can be analyzed in terms of bargaining power. The obsolescing bargain concept refers to the perceived decline in the bargaining position of a multinational enterprise (MNE) in an investment project vis à vis the host country's government. Fagre and Wells (1982), in a study of U.S. multinationals in Latin America, provide some evidence that the bargain struck between a foreign investor and a host country is influenced by the resources brought by the investor and by the number of firms offering similar resources. A crucial resource is a technological advantage. This point is stressed by Ting (1985, 1988). The cost-benefit analysis to identify the worth of (demand for) a project in a particular country can include an evaluation of these variables (Ting, 1988): industry type, number of competing local firms, market share of local firms, project's market share, industry priority, project's technological lead, project's balance of payments contribution, number of MNEs of other nationality, corporate image, and compliance with the country's investment code and regulations. A firm has direct control over some of these variables. Others are inherent in the sociopolitical or economic environments. See Poynter (1982) and LeCraw (1984) for empirical evaluations of bargaining power models.

de la Torre and Neckar's Approach to Political Risk Forecasting

de la Torre and Neckar (1988) approach the task of political risk forecasting from a sociopolitical viewpoint. Their contribution is important because it represents an attempt to combine macro and micro level judgements about risks faced by foreign affiliates of MNEs. Their analytical model has two stages. Stage I is a macro level examination of a series of national characteristics—economic, social and political forces—that may or may not be critical to the issue of political stability. Econometric and expert models can be used at this stage. The results are typical of those produced by commercial risk assessment agencies (e.g., Frost and Sullivan, Data Resources, Inc.). Stage II is the micro level analysis. This consists of evaluating the results of Stage I in the context of a specific project. Here a manager can identify the features of the project that either increase or reduce the possible negative consequences of particular events. Sociopolitical factors monitored in Stage I are divided into two groups: internal and external. These are illustrated in Table 6.1.

TABLE 6.1. Sociopolitical Factors in de la Torre and Neckar's Model

INTERNAL

Composition of population
-ethnolinguistic, religious, tribal or class heterogeneity
-relative shares in economic and political power
-immigration and outmigration

Culture
-underlying cultural values and beliefs
-religious and moral values
-sense of alienation with foreign or modern influences

Government and institutions
-constitutional principles and conflicts
-resilience of national institutions
-role and strength of army, church, parties, press, educational establishment, etc.

Power
-key leaders' background and attitudes
-main beneficiaries of the status quo
-role and power of internal security apparatus

Opposition
-strength, sources of support, effectiveness

General indicators
-level and frequency of strikes
-riots and terrorist acts
-number and treatment of political prisoners
-extent of official corruption

EXTERNAL

Alignments
-international treaties and alignments
-position on international issues, UN voting record

Financial support
-financial aid, food and military assistance
-preferential economic and trade linkages

Regional ties
-border disputes
-external military threat or guerrilla activities
-nearby revolution, political refugees

Attitude towards foreign capital and investment
-national investment codes
-polls of local attitudes towards foreign investors
-court proceedings and disputes

General indicators
-record on human rights
-formal exiled opposition groups
-terrorist acts in third countries
-diplomatic or commercial conflict with home country

Source: Jose de la Torre and David H. Neckar. "Forecasting Political Risks for International Operations" International Journal of Forecasting 4 (1988), 221-241.

APPLYING THE ANALYTICAL FRAMEWORKS TO THE EGYPTIAN EXPERIENCE

Sociopolitical analysis is difficult. The major reason for this is that empirical data on sociopolitical events and their impact on international business are hard to collect (de la Torre and Neckar, 1988). Until recently there has not been sufficient theoretical work that specified causal relationships between political, social and economic data, and the implications for business. The frameworks discussed in this chapter make important contributions to such theoretical work. To judge their applicability to real situations, we illustrate each framework in the context of Egypt's economic liberalization.

Egypt provides a useful case for the study of the sociopolitical aspects of international marketing. Egypt's "infitah" can be used to illustrate how social and political change interacts with economic development to produce policies that have important implications for foreign firms. "El-Infitah" (the opening) is an "open-door" investment policy that was implemented with the passing of Law 43 in 1974. This represented a dramatic shift in Egypt's foreign policy toward the West and followed Sadat's rise to power in mid-1971. Infitah was initially designed to create conditions attractive to foreign investment capital. Law 43 covered five areas with respect to benefits and privileges of foreign investors: taxation; customs duties and import/export regulations; expropriation; exchange control; and business and labor law. Foreign investment was welcomed in the form of joint ventures with the private and/or public sector. All such ventures were free from public sector control. Priority was given to projects designed to generate exports, encourage tourism, or reduce the need to import basic commodities, as well as to projects requiring advanced technical expertise or making use of patents of trademarks therefore promoting technology transfer. Investments were judged on the case-by-case basis and the General Authority for Investment and Free Zones (GAIFZ) was the "control" mechanism or investment-licensing agency. Full details of Law 43 are given in Davies (1984).

Gillespie's Coalition Cycle

Gillespie's coalition cycle was presented in her book (1984) as a model to explain sociopolitical change in Egypt during the Sadat era. One coalition cycle was identified. Rice and Mahmoud (1989) conducted an in-depth examination of the coalition cycle during Mubarak's presidency in the 1980s. Using published statistics, press reports and research studies as information sources, they identified coalition stages in this decade. These are illustrated in Table 6.2. The second coalition cycle began with a return to the development coalition as Mubarak emphasized the original goals of the economic liberalization ("infitah"). Brewer (1983) conjectured that a new political leadership may tend to maintain policy continuity as reassurance to foreign investors. Emphasis was placed on the productive sector rather than the consumer sector. This was because more equitable distribution of Egypt's financial resources was essential. Egypt's new relative wealth (an average annual rate of growth of 8.5% during 1973-84, the period of economic liberalization, as compared to 3.8% during 1965-73) had remained largely inaccessible to the mass of the population (Jabber, 1986). The falling price of oil was also beginning to expose certain weaknesses in infitah (see Moench, 1988, for an analysis).

A capitalist coalition can be identified in the mid 1980s as the government focused on the private sector, maintaining tax incentives and facilitating the acquisition of operating licenses, for example. The coalition cycle, as envisaged by Gillespie, places the emphasis on local investors in a country, the government of that country, and foreign investors. The importance of the influence of outside actors should not be overlooked, however. Inter-nation cooperation and conflict is likely to be significant in the formulation of a governmental policy towards foreign firms (for example, see Schollhammer and Nigh, 1986, and Nigh, 1985). In the case of Egypt inter-nation cooperation has occurred, according to Moench (1988). He explains that funding agencies (such as the International Monetary Fund) and the United States continue to press Egypt to dismantle the huge public sector. The U.S. Embassy *Economic Report* (June 1986) suggests that infitah would be successful if only

TABLE 6.2. The Coalition Cycle in Egypt: An Illustration During Mubarak's Presidency

Coalition	Parties in Coalition	Goal of Coalition	Examples of Events
Development Coalition	Government & Foreign Investors	To promote development that satisfies other political interest groups in the country besides local investor group.	Sadat's assassination in 1981: Mubarak pledges to return infitah to original goals. Emphasis is to be on productive sector rather than consumer sector.
Capitalist Coalition	Foreign & Local Investors	To promote business interests in general against other conflicting interests of the host government.	1985 Emphasis on growth of private sector: tax incentives for private sector firms, facilitating acquisition of operating licenses, encouraging joint ventures.
Nationalist Coalition	Local Investors & Government	Local investors wish to retain government-granted benefits and deny them to foreign investors.	Joint ventures with increasing local participation eg. 75% local manufactured content in Suzuki/Modern Motor Co. venture. Dissatisfaction among various groups: political unrest & rise of opposition groups because of a lack of emphasis on indigenous industry, particularly agriculture and food. Foreign firms hesitant to trade/invest because of political and economic uncertainty.

Source: Gillian Rice and Essam Mahmoud. Economic Liberalization in Egypt: Coalition Cycle Analysis and Implications for International Marketers. In T. Cavusgil (Ed.), Advances in International Marketing Volume III, JAI Press, Inc., 1989.

the "socialist" economic structures left over from the Nasser period could be eliminated. Moench (1988) dubs infitah the "American solution."

Therefore, an improvement in the coalition cycle framework might be the consideration of other external actors (besides foreign investors) as interest groups having influence on or being party to a coalition.

Current government policies suggest that Egypt is in the nationalist phase. Policies are characterized by increasing governmental control over trade and joint ventures. Examples include import controls and increased local participation in joint ventures. Yet these policies are not only a response to pressure on the part of local business to ally with the government to retain or gain more control vis à vis foreign investors. It is not enough to examine sociopolitical forces. The policies are also a response to a dramatic worsening of the Egyptian economy. The economy is burdened by a large trade deficit as well as a chronic foreign exchange shortage. This is exacerbated by returning migrant workers from the Gulf and rapid population growth which result in urban unemployment.

The coalition cycle could be a valuable tool for international marketers. What is needed is further investigation of coalition cycles in other countries and a comparison of the implications for marketers in different countries. Amine (1988) applied the coalition cycle to the Moroccan sociopolitical environment, for example.

Bargaining Power Analysis

It is more difficult to evaluate the bargaining power models and the de la Torre and Neckar forecasting approach because these analytical frameworks have been developed for micro level use by individual firms. Here, the illustration and evaluation are necessarily general.

The bargaining power concept is akin to the coalition cycle in some respects because it entails judging the advantages of one firm (particularly the technological advantage) vis à vis those of other firms, including Egyptian firms. In Egypt, as in other developing and industrializing countries, one goal of economic liberalization was to gain technology transfer. Hence foreign firms with technology to offer were in a strong bargaining position. It appears that this led to foreign firms' being able to enter and operate in Egypt on very favorable terms using primarily Egyptian public and private capital. Moench (1988) reports that the injection of new capital from outside Egypt was almost negligible. Foreign firms approved by the government were able to claim as "capital" their names, market share and expertise. Moench also comments that the few firms with industrial projects were capitalized at such low levels that their scale was simply not sufficient for any substantial technology transfer. Investments in land reclamation and food industries were minor as compared to those in finance, construction and building materials (Najafbagy, 1985). Most investment has been in commercial and service activities.

Part of this failure of infitah to control for the investments and benefits received from foreign firms can be attributed to the lack of organization in the government. Ayubi (1982), for example, noted that the performance of the Egyptian bureaucracy was declining sharply in quality, when the desire to encourage foreign investment was actually calling for a more innovative, flexible and efficient bureaucracy. The problem may be attributable to the massive state bureaucracy established in the mid-1960s (Gillespie, 1984). This emphasizes the importance of an historical perspective in sociopolitical analysis.

Ting (1988) distinguishes between various "entry management systems" used by nations for the control of foreign investors. These systems can structure the bargaining process between firms and a host country and contribute to a host country's success in achieving goals. Therefore careful understanding of the systems can help a marketing manager obtain the best entry terms for the firm. Ting identifies countries with institutionalized entry management systems (EMSs) such as Singapore, Taiwan and South Korea; countries with less ambitious goals and poorly defined and inefficiently implemented industrialization objectives that tend to have little or no formally organized EMSs; and industrialized countries such as the U.K. or France that have EMS-type institutions for the revitalization of their economies. Egypt exemplifies the middle category as a country with a poorly organized EMS. The principal criticism made by investors is the lack of coordination between various government agencies and their apparent inability to honor important undertakings. For example, while Egypt has an investment authority (GAIFZ), a customs exemption from GAIFZ does not automatically mean that the customs service will abide by it. Egypt is making efforts to centralize the investment approval process, however (*Middle East and African Economist*, 1982).

Despite a rather weak position on the part of the Egyptian government because of policy instability and bureaucratic inertia, a foreign firm should not become complacent. Competition between foreign investors remains strong, as witnessed by the recent bidding by major automobile firms for a joint venture opportunity with the Egyptian public sector auto manufacturer and thus a strong foothold in a growing market (for example, see *Wall Street Journal*, March 6, 1986). Foreign investors must be willing to undertake joint ventures, maximize the export intensity of the project, maintain the project's technological contribution, use local inputs to the production process and be socially responsible in order to maintain a strong bargaining position (Ting, 1988).

de la Torre and Neckar's Approach to Political Risk Forecasting

As with the evaluation of bargaining power models, the evaluation of the political risk forecasting approach is limited to a country (macro) level analysis. In the context of de la Torre and Neckar's framework, key internal sociopolitical forces in Egypt include outmigration, religious values and the nature of the opposition. Major external forces are the close ties with the U.S., infitah and local attitudes towards foreign investors (refer to Table 6.1). Of course, in the complete analysis advocated by de la Torre and Neckar, economic factors are examined, in addition to the project-related factors for the micro level analysis. Here, we focus only on the sociopolitical aspects of their framework.

Egypt has a large and diversified work force which represents one of the country's most valuable resources for long-term development. The work force is also the source of a number of problems that appear to have been caused by government mis-management. Low productivity is endemic, particularly in the large public sector, because of over-staffing policies as short-term solutions to unemployment. The Egyptian government encouraged outmigration to the Gulf states and Libya during the oil boom years. This was not without cost to the Egyptian economy. Although workers' remittances were an important source of revenue, these have fallen dramatically as the oil price fell and in 1986 when thousands of Egyptian workers returned home (Moench, 1988). Moench provides a detailed analysis of how migration led to production losses in Egypt and to the attrition of agricultural land because of trained labor shortages at home. Furthermore, he writes "the Egyptian workers sojourning in the oil-producing countries return demoralized by their treatment, devoid of whatever proletarian consciousness they might have acquired before migrating, and far less interested in improving working conditions in Egypt than in discovering petty entrepreneurial ways to make their wealth grow without working" (p. 180).

According to Hatem's (1988) study of the Egyptian middle class, those who migrated to the Gulf, along with foreign-educated members of the middle class and key government bureaucrats, are comfortable with secular and Western lifestyles. These are the liberal elements of the middle class. In contrast, the stagnation of the public sector and worsening economic conditions created difficulties for some segments of the lower middle class. These people have used Islamicism to question the impact of infitah on the Islamic character of society. Moench (1988) identifies this as a shift to the "cultural right." This is a term to mean that these people see social (and economic) problems as moral problems, asserting the causal priority of the latter. The national debate in Egypt thus centers on religious versus secular solutions to the country's problems.

The major question for international marketing managers is what impact these social changes might have on political views, and subsequently, on policy changes. While it is true that the opposition is extremely fragmented (Moench, 1988; Jabber, 1988), one example of a policy reorientation is the strong government campaign to encourage people to buy Egyptian products (*South*, February 1988).

Nevertheless, external forces (most notably the U.S. and the International Monetary Fund) remain strong and infitah has not yet been abandoned or changed fundamentally. Despite growth in nationalism, the open-door policy is still supported. A "new class" has arisen as a result of the policy and this group provides a base of support for the government. Jabber (1986) explains that although this group is relatively small, it has accumulated much economic and political power. This group consists mainly of entrepreneurs, professionals and high-salaried employees of the private economy. Support also comes from the liberal elements of the middle class identified by Hatem (1988).

This brief review of some aspects of the Egyptian sociopolitical environment in the context of de la Torre and Neckar's approach shows that the monitoring of a series of variables and the identification of a number of key factors can be useful. de la Torre and Neckar do not provide weights for the factors. For the approach to be valuable for micro level analysis, firms should apply their own weighting scheme. This will vary across countries. de la Torre and Neckar have not addressed the variable measurement issue, even though they mention that econometric models could be used in the analysis. While this may be possible, the factors might be analyzed to more advantage using narrative and in the context of a systematic scenario (for details, see Ascher, 1982).

CONCLUSIONS AND RECOMMENDATIONS FOR FUTURE DEVELOPMENTS IN SOCIOPOLITICAL ANALYSIS

Each of the three analytical frameworks examined in this chapter has something to offer the international marketer seeking to understand the sociopolitical environment. They also suffer from a number of weaknesses. In Table 6.3 we provide a summary evaluation.

The application of the frameworks to the case of Egypt's infitah highlights the fact that one cannot examine sociopolitical forces in isolation. These forces interact in important ways with economic performance. For example, the magnitude and effects of change in Egypt are closely related to which population groups benefit from the results of government economic policies and to what degree

TABLE 6.3. Summary Evaluation of Three Sociopolitical Analytical Frameworks

Framework	Advantages	Weaknesses
1. Coalition cycle (Gillespie, 1984)	Dynamic. Explicit in causality from coalition formation to policy implementation.	Difficult to distinguish between stages because of overlap. Focus on only three interest groups. Needs adapting for micro level use. Measurement problems.
2. Bargaining power (Vernon, 1971, 1977; Ting, 1985, 1988)	Dynamic. Shown to be applicable across different time periods and countries. Micro level.	Identification and measurement of key variables. Information collection (competitive intelligence).
3. Political risk forecasting (de la Torre and Neckar, 1988)	Comprehensive. Complete framework is for micro level analysis. Good structure for monitoring situation in a host country.	Provides listing of factors only. Does not suggest interactions, causal relations, policy impacts. Measurement problems.

those population groups that do not benefit can exert pressure to change policy.

The Egyptian illustration also emphasizes the need to go beyond monitoring sociopolitical variables on a "checklist" basis. As Ascher (1982) advocates, it is essential to understand the dynamics of sociopolitical processes. For example, an important weakness of the de la Torre and Neckar framework is that although it includes a comprehensive listing of sociopolitical variables to be monitored, it does not suggest how these variables might be inter-related or impact government policy. In contrast, Gillespie's (1984) explanation of the coalition cycle goes into considerable detail of the reasons why coalitions form. This is important because understanding the process is helpful to predicting a new coalition and what the implications for international marketing might be.

Yet with substantial effort on the part of management, all three frameworks can be adapted for use as predictive tools and for gauging the impact on specific projects. That such an effort is generally required is a practical problem. This was acknowledged by de la Torre and Neckar (1988). Sociopolitical analysis, because of its attention to detail, requires many managerial and financial resources. The extent of efforts made has to be weighed against the likely success of the analysis in contributing to improved marketing strategies.

Resource demands are exacerbated by measurement problems. For example, in the bargaining power models, it is first necessary to identify what provides the strength of the parties (the host government and the MNE) to a "bargain." Are they able to maintain these strengths? Measurement of the identified variables has to be subjective at best. Management must also decide how to estimate the relative technological strength, for example, of the host country domestic firms and other foreign firms. In the coalition cycle model, measurement difficulties also occur. A specific and unresolved issue is how to identify precise cut-off points for the various coalition stages as there are inevitable overlaps.

Despite these difficulties, sociopolitical analytical frameworks provide a systematic basis for evaluating environments of foreign countries. Systematic approaches facilitate managerial decision making. These approaches should be used judiciously, however, in order to obtain the greatest benefit. For example, keys to the successful use of sociopolitical analysis for international marketing include the following.

1. Use the analysis to understand the dynamics of sociopolitical processes that might lead to relevant policy changes.
2. Conduct the analysis at multiple levels (macro and micro). All three frameworks discussed here could be used in this way. It is crucial to ensure that firm or project specific implications are made explicit. Only then can the analysis be valuable for decision making. For example, what does the analysis imply about the market entry strategy a firm should follow? What creative advertising strategy would be appropriate, given renewed social orientation towards Islamic values in Egypt?
3. Use the analysis in a proactive rather than a reactive way. Unfortunately most firms tend to be reactive in their response

to sociopolitical change (Kobrin, 1982). One reason may be the inadequacy of analytical frameworks in helping firms be proactive. The key is to understand the situation so well that one can anticipate processes and events. To this end, more theoretical as well as empirical research is needed to clarify the relationships between sociopolitical variables and government policy outcomes relevant to international marketing.

Despite the need for improvements, the frameworks discussed in this chapter can make a useful contribution to an international marketing manager's attempts to understand the sociopolitical environment and formulate strategies accordingly.

REFERENCES

Amine, Lyn S. (1988). Changes in the Marketing Environment in Morocco. Paper presented at the annual meeting of the Academy of Marketing Science, Montreal, Canada.

Ascher, William (1982). Political Forecasting: The Missing Link. *Journal of Forecasting*, Vol. 1, pp. 227-239.

Austin, James E. & Yoffie, David B. (1984). Political Forecasting as a Management Tool. *Journal of Forecasting*, Vol. 3, pp. 395-408.

Ayubi, N.M. (1982). Bureaucratic Inflation and Administrative Inefficiency: The Deadlock in Egyptian Administration. *Middle Eastern Studies*, Vol. 18, No. 3, pp. 186-229.

Brewer, Thomas L. (1983). The Instability of Governments and the Instability of Controls on Funds Transfers by Multinational Enterprises: Implications for Political Risk Analysis. *Journal of International Business Studies,* Vol. 3, pp. 147-157.

Brewer, Thomas L. (1985). Political Risks in the Future. In Brewer, T.L. (ed.), *Political Risks in International Business*. New York: Praeger Publishers, pp. 337-349.

Davies, M.H. (1984). *Business Law in Egypt*. Deventer, The Netherlands: Kluwer Law and Tax Publishers.

de la Torre, Jose & Neckar, David H. (1988) Forecasting Political Risks for International Operations. *International Journal of Forecasting*, Vol. 4, pp. 221-241.

Economic Report (1986). U.S. Embassy in Egypt.

Fagre, Nathan & Wells, Louis T., Jr. (1982) Bargaining Power of Multinationals and Host Governments. *Journal of International Business Studies*, Vol. 13, No. 2, pp. 9-23.

Gillespie, Kate (1984). *The Tripartite Relationship Government, Foreign Inves-*

tors, and Local Investors During Egypt's Economic Opening. New York: Praeger Publishers.

Hatem, Mervat (1988). Egypt's Middle Class in Crisis: The Sexual Division of Labor. *Middle East Journal*, Vol. 42 No. 3, pp. 407-422.

Jabber, P. (1986). Egypt's Crisis, America's Dilemma, *Foreign Affairs*, Vol. 64, No. 4-5, pp. 961-980.

Jabber, P. (1988). Forces of Change in the Middle East. *The Middle East Journal*, Vol. 42, No. 1, pp. 7-15.

Kobrin, Stephen J. (1982). *Managing Political Risk Assessment*. Berkeley and Los Angeles: University of California Press.

Lecraw, Donald J. (1984). Bargaining Power, Ownership, and Profitability of Subsidiaries of Transnational Corporations in Developing Countries. *Journal of International Business Studies*, Vol. 15, No. 1, pp. 27-44.

Middle East and African Economist (1982). Vol. 36, No. 7-8, p. 43.

Moench, Richard U. (1988) Oil, Ideology and State Autonomy in Egypt. *Arab Studies Quarterly*, Vol. 10, No.2, pp. 176-192.

Najafbagy, R. (1985). Operations of Multinational Corporations and Local Enterprises in Arab Countries. *Management International Review*, Vol. 25, No. 4, pp. 46-57.

Nigh, Douglas (1985). The Effect of Political Events on U.S. Direct Foreign Investment: A Pooled Time-Series Cross-Sectional Analysis. *Journal of International Business Studies*, Vol. 16, pp. 1-17.

Poynter, Thomas A. (1982). Government Intervention in Less Developed Countries: The Experience of Multinational Companies. *Journal of International Business Studies*, Vol. 13, No. 1, pp. 9-26.

Rice, Gillian & Mahmoud, Essam (1989). Economic Liberalization in Egypt: Coalition Cycle Analysis and Implications for International Marketers. In Cavusgil, T. (ed.), *Advances in International Marketing Volume III*, Greenwich, CT: JAI Press, Inc.

Robock, Stefan H. (1971). Political Risk: Identification and Assessment. *Columbia Journal of World Business*, July-Aug., pp. 6-20.

South (1988). Infitah Injects the Fizz. February, 9.

Ting, W. (1985). *Business and Technological Dynamics in Newly Industrializing Asia*. Westport, CT: Quorum Books.

Ting, W. (1988). *Multinational Risk Assessment and Management*. Westport, CT: Quorum Books.

Vernon, Raymond (1971). *Sovereignty at Bay*. New York: Basic Books.

Vernon, Raymond (1977). *Storm over the Multinationals: The Real Issues*, Cambridge, MA: Harvard University Press.

Wall Street Journal (1986). General Motors to Assemble Cars in Egypt. March 6.

Chapter 7

Country of Origin Stereotyping: The Social Linkage Effect Upon Foreign Product Attributes

Frank Renwick
Rebecca M. Renwick

Stereotypes developed as a result of a product's country of origin have been referred to as country of origin stereotyping. This has also been described as a country of origin effect and has been shown to influence both consumers' and importers' perceptions and evaluation of foreign products. Social linkages with nationals of a foreign country have previously been shown to influence the overall preference for foreign products when purchasing experience has had no effect on the overall stereotype. The research reported here investigated which product attributes were most affected by Caribbean importers' social linkages. The product attributes examined in this study included quality, service, delivery, availability, and price. These were addressed for eight exporting nations: Canada, France, Italy, Japan, Netherlands, United Kingdom, United States, and West Germany.

Recent research has indicated that importers' stereotypes concerning an exporting country are positively influenced by friendships (social linkages) with individuals who are nationals of that country. These findings suggest that countries which are now home to a higher percentage of an exporting country's nationals constitute favorable export target markets. However, this work examined country of origin stereotyping at only an overall preference level of analysis. The present research examines stereotyping at a product attribute level of analysis and has implications for a product strategy approach to targeted export markets as well as for the selection of export products and markets based upon perceived product attributes.

CONCEPTUAL FRAMEWORK

Country of Origin

The country of origin stereotyping research with respect to both the industrial and consumer markets has indicated significant effects of the country of origin upon buyers' evaluations of products (Bilkey and Ness, 1982). These stereotyping effects have been demonstrated at the generalized product level (Gaedeke, 1973; Lillis and Narayana, 1974; Nagashima, 1970; Reierson, 1966), the product class level (Gaedeke, 1973; Kaynak and Cavusgil, 1983; Nagashima, 1970; Nagashima, 1977), the product form, and product item levels (Etzel and Walker, 1974; Gaedeke, 1973; Hakansson and Woots, 1975; Heslop, Liefeld, and Wall, 1987; Johansson, Douglas, and Nonaka, 1985; White and Cundiff, 1978).

Most of the research in this area has focused upon the consumer markets since these are easier to access than the industrial markets. However, the industrial market research has shown congruence with that of the consumer markets in terms of the stereotyping effect of the country of origin upon product evaluations. This effect has been demonstrated at the generalized product level (Cattin, Jolibert, and Lohnes, 1982; Jaffe and Nebenzahl, 1984; Nagashima, 1977; Renwick and Renwick, 1988; White, 1979), the product class level (Nagashima, 1977) and the product item level (Angelmar and Pras, 1984; Hakansson and Wootz, 1975; White and Cundiff, 1978) for industrial products.

Stereotyping

Lippman (1922) suggested that the real environment is too large and complex to deal with. He postulated that individuals are not equipped to handle excess information and, as a result, they reconstruct reality in a more simplistic form which allows them to come to grips with their environment. Lippman (1922) viewed this reconstruction and simplification process, or stereotyping, as a normal process through which individuals become able to cope with their environments. These stereotypes or "pictures in our heads" appeared to possess both cognitive and affective dimensions. Research has shown that the most powerful stereotypes are those of race, sex, and abnormalities (Gilman, 1985).

Ethnic and racial stereotyping have been examined by Allport (1958). He identified implicit behavioral theories which were based upon the explicit differences in physical appearances evident in different races and ethnic groups. This kind of stereotyping is trait-driven in that the objects of this stereotyping are ascribed traits which are presumed to prescribe their behaviors. This tendency to personalize stereotypes identified by Allport (1958) has received broad support (Campbell, 1967; Heider, 1958; Inkeles and Levinson, 1968; Nisbett and Ross, 1980; Tajfel, 1969).

The outcomes of stereotyping are as diverse as would be expected from a process with different applications. However, there has been overall agreement that stereotyping involves social perception and attribution. The social perception is usually linked to a cue or stimulus and the attribution is normally dispositional rather than situational. These social perceptions and attributions are exemplified by the stereotypes which are held of foreigners. These images form the basis for how individuals may feel and act toward the nationals of other countries and, in fact, there is usually substantive agreement on the set of attributes inherent in this kind of stereotype (Mackie, 1973; Secord and Backman, 1974). This form of stereotyping has been explained in terms of ethnocentrism (Brewer, 1979; Levine and Campbell, 1972).

When a stereotypical trait is associated with a group, the members of that group become identified with that trait even though the manifestation of the trait may vary considerably among the members of the group (Tajfel, Sheikh, and Gardner, 1964). Polarization and neutralization are also part of the process of stereotyping (Taylor, Fiske, Etcoff, and Ruderman, 1978). Polarization is the amplification of the differences between stereotypical categories and neutralization involves minimization of the similarities between those same stereotypical categories. The attributions associated with a stereotypical category normally surpass the criterion attribution, which was used to classify the individuals or objects into a stereotypical category (Asch, 1946). The dispositional traits ascribed by the stereotype not only exceed the observable criteria responsible for the categorization, but may also be part of a trait set about which individuals hold implicit theories (Nisbett and Ross, 1980). These are evident in the common descriptive sets which are used for most stereotypes (e.g., old miser, old hermit, young rebel, etc.).

Stereotyping is a memory-driven process so that categories which have a high frequency of use are more available in memory than those with a low frequency of use (Tversky and Kahnemann, 1973). As the occasion arises, memory load appears to influence the cognitive rewriting of stereotypical categories. A high memory load situation is one in which multiple exposures to nationals of a foreign country cause the stereotype to be continually updated on the basis of new information. A low memory load situation is one in which limited exposure to one or two nationals of a foreign country gives rise to the stereotypical category (Rothbart, Fulero, Jensen, Howard, and Birrell, 1978).

Cognitive consistency or the selective retention of experiences or perceptions which are consistent with our stereotypical categorization has been described by Cohen (1977). Further, Cantor and Mischel (1977) have suggested that individuals enrich or enhance their stereotypes more on the basis of implicit attribution than observed behavior. Bartlett (1932) has shown that memories are apt to be more coherent than the actual events. In fact, it appears that individuals bend and alter details to make them fit with their existing picture of reality. By structuring the ambiguous portions of their perceptions in this manner individuals perpetuate their stereotypes (Tajfel, 1969). Jones (1982) has also identified the tendency to ignore new information when dealing with stereotypical categories. This is particularly evident when the stereotype is negative in nature (Briscoe, Woodward, and Shaw, 1967).

Once an image or stereotype of a culture, country, foreign nationals, or their products has been developed, these beliefs will partially determine an individual's affective orientation and may be the main determinant of intended and subsequent behaviors (Fishbein and Ajzen, 1975).

Knowledge

The degree of preference for foreign countries has been shown to vary with the nature of knowledge and knowledge levels concerning those countries (Grace, 1954; Grace and Neuhaus, 1952). Those countries about which individuals had little knowledge were either disliked or affectively neutral while the more highly preferred countries were those about which individuals had more extensive knowledge. The nature of this knowledge was also an influential factor, that is, positive knowledge concerning a country resulted in a preference for it. Similarly, liking for a foreign national or having one for a friend resulted in a positive orientation towards his country of origin and provided an opportunity to learn more about it.

Social Linkages

Social linkages are pursued where there are high levels of attraction. Individuals are more attracted to others who are similar to themselves and to others who are familiar (Hill and Still, 1981; Lott and Lott, 1972). Specifically, attraction is associated with similarities in beliefs and values (Hill and Still, 1981). Further, when individuals form friendships on the basis of familiarity, similarity to one another is typically assumed (Moreland and Zajonc, 1982). In addition, positive interpersonal relations with foreign nationals will give rise to favorable stereotypes (Amir, 1969; Zajonc, 1968) and, conversely, negative interpersonal relations with foreign nationals will elicit negative stereotyping (Salter and Teger, 1975; Sell, 1983).

Levels of Abstraction

Degree of specificity, or the level of abstraction at which people have knowledge about product attributes varies quite widely (Hirschman, 1980). Three levels have been identified: (a) the abstract, multidimensional level (b) the concrete, specific level and (c) the less abstract, more objective category which lies between the other two (Geistfeld, Sproules, and Badenhop, 1977). These distinct levels of meaning have been associated with subsets of attributes. Specifically, Peter and Olson (1987) have described the abstract level (A) as having the largest multidimensional set while the concrete level (C) was unidimensional and the in-between category (B) had a small multidimensional set.

Collectively, the product attributes examined in the context of the research reported here represented all three of these levels of abstraction. Specifically, level A was addressed in terms of quality, level C in terms of delivery and price, and level B in terms of service and availability.

RATIONALE

Stereotyping research has identified racial stereotyping as one of the most powerful stereotypes, ranking alongside gender and psychiatric disorder stereotypes (Gilman, 1985). The **country of origin effect** is principally an example of racial stereotyping. This type of stereotyping is trait-driven or attributed to disposition rather than to situations (Allport, 1958). There is also evidence to suggest that part of the country of origin effect can be attributed to the environment, that is, climate, geography, political environment, and the level of economic development. However, the major orientation appears to be one of a national stereotype for which there is usually wide agreement (MacKie, 1973; Secord and Backman, 1974).

Social linkage effects upon country of origin stereotypes are easy to understand when considered in terms of knowledge and exposure. If contact with foreign nationals is limited, then the stereotypical category is formed as a result of a low memory load situation. But, if there are multiple exposures to foreign nationals, as in the context of a friendship, then the stereotype is continually updated and this constitutes a high memory load situation (Rothbart et al., 1978). Further, friendships (social linkages) are usually based upon attraction and also provide some benefit to the participants (Amir, 1969; Hill and Stull, 1981; LaGaipa, 1977; Lott and Lott, 1972; Moreland and Zajonc, 1982; Renwick, 1985; Sell, 1983; Zajonc, 1968). Thus, friendships engender favorable associations which can positively affect the stereotype (Renwick and Renwick, 1988). During the course of friendship, knowledge levels concerning the friend's culture, country, or lifestyle increase and, therefore, it might be expected that these higher knowledge levels will also be associated with increased preference levels for the home country of the friend (Grace, 1954; Grace and Neuhaus, 1952).

Product attributes might be arranged upon a level of abstraction continuum which ranges from the abstract, multidimensional to the concrete, unidimensional (i.e., from quality through service and availability to delivery and price). A stereotypical category arising out of the implicit attribution process would belong at the abstract end of the scale as would the country of origin stereotype. The robustness of the country of origin stereotype has been attributed to the cognitive consistency theory (Bartlett, 1932; Briscoe et al., 1967; Cantor and Mischel, 1977; Cohen, 1977; Jones, 1982; Tajfel, 1969).

RESEARCH QUESTIONS

Previously reported research examined the effects of purchasing experiences and social linkages upon cross-cultural stereotyping (Renwick and Renwick, 1988). In this research, importers' purchasing experiences were examined to determine if there were any differences in the country of origin perceptions held by those with specific purchasing experiences with a country as compared with those who had no such experiences. Their social linkages were also examined to determine if there were any differences in their country of origin perceptions based upon the nationalities of their friends. The combination of purchasing experiences and social linkages was also examined.

The findings indicated that importers were more likely to buy from countries with which they had social linkages. Purchasing experiences did not appear to significantly affect the country of origin stereotypes. In general, social linkages appeared to have a strong positive influence upon the country of origin stereotype regardless of the purchasing experiences with that country. The important findings of this research were that importers were more likely to buy from those countries with whose nationals they had formed social linkages than from those countries with whose nationals they had formed no friendships. The next most important finding was that country of origin stereotyping was robust in that it did not appear to change as a result of purchasing experiences with that country. However, it did change as a function of friendships formed with nationals of that country.

Although social linkages have been shown to strongly and positively influence the country of origin stereotype in terms of a general product preference (Renwick and Renwick, 1988), it has not been determined whether this preference extends to all of the salient product attributes. Etzel and Walker (1974) examined the congruence between attitudes towards foreign products at different levels of abstraction. The levels of abstraction were limited to the product category. A generalized product and specific types of products were examined and the findings indicated that consumers did perceive a difference between the generalized and specific product items.

In the current study, it was expected that there would also be differences based upon the levels of abstraction of the salient product attributes. Specifically, the product attributes most affected by social linkages were expected to be at the abstract end of the levels of abstraction continuum. Those product attributes at the concrete end were expected to evince little or no effect of social linkages.

The country of origin stereotype was measured for product attributes with respect to eight exporting nations who represented the major trading partners of the Caribbean sample. These trading partners are Canada, France, Italy, Japan, Netherlands, United Kingdom, United States, and West Germany.

METHOD

Instrument

Most of the country of origin stereotyping research has used some version of the Nagashima (1970) scales. The original Nagashima instrument used 20 scale items in a semantic differential format. Subsequent research has consisted of reduced versions of these scales augmented with individual variations (Jaffe and Nebenzahl, 1984; White, 1979). Factor analysis of the original scales has produced mixed results varying from two factors accounting for 67% of the variance (Jaffe and Nebenzahl, 1984), to five factors which accounted for 42% of the variance (Naranya, 1981), to no major set of factors (Cattin et al., 1982). These inconclusive results, when combined with the lack of correspondence between

Nagashima's scale items and dimensions (Jaffe and Nebenzahl, 1984), suggested a construct validity problem with many of these country of origin data collection instruments.

Jaffe and Nebenzahl (1984) examined the internal consistency of the Nagashima scales and found coefficient alphas ranging from 0.63 to 0.73, which indicated relatively low internal reliability. The construct validity and reliability problem argued against the use of the Nagashima scales. Therefore, an elicitation process was employed to identify the salient attributes at the product, company, and country of origin levels, as described by Renwick (1988). These attributes had construct validity since they were elicited from the sample population and the reported internal reliabilities ranged from .83 to .93. The five scale items identified were quality, price, availability, delivery, and service.

Vendor selection research has identified price, quality, and service as the salient criteria (Hill, Alexander, and Cross, 1975; White, 1979). Quality, delivery, price, reliability, and service were identified as the common attributes important to both American and Indian buyers in vendor selection (Rao, Rosenberg, and White, 1988). Price, delivery, and quality were identified as the important attributes in vendor bid characteristics (Hakanasson and Wootz, 1975). This research was in substantive agreement with the five scale items which were established by Renwick's (1988) elicitation process. Therefore, these elicited items were incorporated into a data collection instrument which was employed in the research reported here.

Five-position, itemized rating scales for each of the salient attributes and an attribute processing (Q2) question format (Batsell and Wind, 1980; Jaffe and Nebenzahl, 1984) were employed. The level of abstraction used for the product evaluations was purposefully high in order to allow for a wide sectoral representation in the sample. The generalized product format was utilized on the assumption that respondents answered questions about general products based upon their own experiences (Etzel and Walker, 1974; Lillis and Naranya, 1974; Reierson, 1966; Wang and Lamb, 1983). The instrument also included questions regarding their social linkages with individuals from the eight exporting countries identified earlier. The instrument was pre-tested with 30 importers and no difficulties were encountered.

Sample and Procedure

A cross-cultural sample of 600 respondents was systematically drawn from a sample frame of importers in five Caribbean countries, namely, Bahamas, Barbados, Haiti, Jamaica, and Trinidad. The sample frame was compiled from master lists supplied by customs, government ministries (i.e., finance, industry, and development), manufacturers' associations, independent businessmen's associations, and chambers of commerce. Table 7.1 presents the complete distribution of the population sample based upon Standard Industrial Classifications. The grouping of respondents according to sector is shown in Table 7.2.

TABLE 7.1. Caribbean Sample Distribution

Type of Business	Number of Respondents	Type of Business	Number of Respondents
Agriculture & Forestry -		Wholesale -	
Agricultural	1	Farm products	2
Agricultural service	8	Food, beverage, drug, tobacco	20
Logging	1	Apparel, dry goods	10
Manufacturing - food	19	Household goods	8
Beverage	10	Autos, parts, accessories	29
Tobacco products	3	Metals, hardware, building materials	28
Rubber products	2	Machinery, equipment, suppliers	3
Plastic products	6	Other Products	7
Leather and Allied products	3	Retail - Food, beverage, drug	12
Primary textiles	3	Shoe apparel, fabric, yarn	33
Textile products	7	Household furniture, appliances	24
Clothing	21	Auto parts sales and service	22
Wood	5	General merchandise	4
Furniture and Fixtures	14	Other	70
Paper and Allied products	11	Financial - Deposit accepting	4
Printing and Publishing	11	Insurance	1
Primary Metal	1	Other financial	8
Fabricated Metal products	16	Service - Real estate	3
Machinery	4	Insurance, real estate	1
Transportation equipment	1	Business services	2
Electrical and Electronic	15	Health, social service	1
Non-metallic Mineral products	11	Accommodation	8
Refined petroleum and coal	1	Food, beverage	5
Chemicals & chemical products	21	Amusement, recreation	5
Other manufactured products	19	Personal and household	16
Construction - Building,			
General Contracting	3		
Industrial, Heavy Engineering	8		
Trade Contracting	10		
Construction service	6		
Transportation Utilities and			
Communication-Transportation	6		
Storage and warehousing	1		
Communication	1		
Other Utilities	1		

TABLE 7.2. Sectoral Representation of Sample

Sector	Number of Respondents
Agriculture and Forestry	10
Manufacturing	204
Construction	27
Transport, Utilities, and Communication	9
Wholesale	101
Retail	162
Financial	13
Services	41

The products which were represented in this study reflected very diverse importing experiences (Etzel and Walker, 1974). The primary products imported by the Caribbean sample are shown in Table 7.3. The groupings of these products in terms of the Standard Industrial Trade Classifications are presented in Table 7.4.

The survey was conducted by experienced field interviewers who were nationals of each of the five Caribbean countries from which the sample was drawn. Each interview was conducted with the individual (within each company sampled) who was responsible for sourcing imports for that company.

RESULTS

The effect of social linkages upon the perceived product attributes of products imported from the United States, Canada, United Kingdom, Netherlands, Italy, West Germany, France, and Japan was tested by a series of t-tests. To remove the effect of purchasing experience the sample was divided into two groups, those with foreign country purchasing experiences and no foreign social linkages and those with foreign purchasing experiences as well as foreign social linkages. The product attributes for each country were examined in this fashion.

TABLE 7.3. Primary Imports for Caribbean Sample

Imported Products	Number of Respondents	Imported Products	Number of Respondents
Meats	9	Manufactured goods by material	3
Dairy products, eggs	3	Leather, leather manufactures, fur dressed	3
Fish, crustaceans and molluscs	1	Rubber manufactures	4
Cereals	4	Cork, wood manufactures	3
Vegetables, fruits	4	Paper, paperboard, paper pulp board	23
Coffee, tea, cocoa, spices	4	Textile yarn and fabrics	41
Animal feeds	2	Non metallic mineral manufactures	1
Miscellaneous edible products	19	Iron and steel	6
Beverages	5	Non-ferrous metals	10
Tobacco	3	Metal manufactures	16
Crude materials inedible	1	Machinery and transport equipment	2
Cork, wood	11	Power generating machinery	2
Pulp, waste paper	1	Machinery (specialized)	14
Textile fibres and their waste	5	Metal working machinery	2
Crude fertilizers, crude materials	2	General industrial machinery	7
Metal ores and metal scrap	1	Office machines, automated data processing equipment	16
Crude animal and vegetable materials	2	Telecommunications, sound recordings	11
Petroleum and products	2	Electronic appliances, apparatus, machines	40
Animal and vegetable oils, fats, and waxes	1	Road vehicles, parts	37
Fixed vegetable oils and fats	1	Other transport equipment	11
Animal, and vegetable oils, fats, and waxes	5	Miscellaneous manufacturers, products	16
Chemicals and related products	13	Sanitary plumbing, and heating, lighting fixtures	7
Organic chemicals	3	Furniture and parts	13
Inorganic chemicals	7	Apparel and clothing accessories	24
Dyeing, tanning and coloring materials	1	Footware	16
Medicinal and parmaceuticals	19	Professional, scientific, control instruments	3
Essential oils, perfume material, toilet cleansing materials	6	Photographic, optical goods, and watches	7
Artificial resins, plastics, ether, cellulose	15	Miscellaneous manufactured articles	47
Chemical materials and products	4	Commodities, transactions, other	5

As demonstrated in Table 7.5, results of the t-tests indicate that social linkages do affect perceptions concerning the quality of foreign products. For seven of the eight countries, the product **quality** means for the group with social linkages are greater than for the group without social linkages; however, there is a significant difference for only five of the eight countries. These countries are the United States, Canada, United Kingdom, Italy, and West Germany. For six of the eight countries, the **service** attribute means are higher for importers with social linkages; however, significant differences emerged for only three (i.e., Canada, United Kingdom, and West

TABLE 7.4. Product Groupings by Standard Industrial Trade Classifications (SITC)

SITC	Number of Products
Food, live animals	46
Beverages, tobacco	8
Crude materials, inedibles	23
Minerals, fuels, lubricants	2
Animal and vegetable oils and fats	7
Chemicals	68
Manufactures by material	110
Machinery, transport equipment	141
Manufactures, articles	143
Commodities, transactions	5

TABLE 7.5. Product Attributes: t-test Results

	t-values				
Country	Quality	Service	Availability	Delivery	Price
U.S.A.	4.66**	1.58	0.23	0.77	0.33
Canada	6.35**	2.28*	2.88**	0.91	0.73
United Kingdom	3.00**	2.96**	1.05	1.96	1.02
Netherlands	1.62	0.58	1.15	0.70	0.36
Italy	3.61**	1.18	2.50*	1.80	0.68
West Germany	3.21**	2.61*	2.42*	1.01	0.71
France	0.15	1.36	1.95	1.18	0.62
Japan	0.47	0.55	1.44	0.46	2.03

$^{*}p < .05$. $^{**}p < .01$.

Germany). The **availability** attribute means were higher for those with social linkages for seven countries, but these differences were significant only for Canada, Italy, and West Germany. The means for the **delivery** attribute were higher for importers with social linkages for three exporting countries but no significant differences emerged. The **price** attribute means for the two groups of importers did not differ significantly for any of the eight exporting countries.

These results indicate that quality is the product attribute most affected by social linkages with an exporting nation. There is a smaller effect on service and availability but no effect on price or delivery. These findings are consistent with the abstract-concrete continuum of product attributes discussed earlier. They are also congruent with the similarities between stereotyping and the levels of abstraction of the generalized foreign product attributes as perceived by importers. There is also substantive agreement that the perception of quality of foreign products is affected by the country of origin stereotype (Etzel and Walker, 1974; Gaedeke, 1973; Heslop et al., 1987; Kincaid, 1970; Lillis and Narayana, 1974; Reierson, 1966; Schooler, 1971; White and Cundiff, 1978).

DISCUSSION AND CONCLUSIONS

Previous research shows that positive social linkages (friendships) have a strong, favorable influence upon importers' country of origin preferences (Renwick and Renwick, 1988). Further, these stereotypes are not significantly altered even after purchasing experience with that country, but are altered as a result of the friendships formed with nationals of that country. When examined at a product attribute level, for those importers who have had purchasing experience, social linkages with the foreign nationals of those exporting nations positively influence the perceived quality of products from those nations. The important distinction is that **even after purchasing experience** with that country the perceived quality of the products from that country are higher if social linkages existed with nationals of that country.

The success of Japanese and American manufacturers in international markets is attributed largely to product quality (Garvin, 1984; Ouchi, 1981). Phillips, Chang, and Buzzell (1983) state that there is a significant positive influence of perceived product quality upon market share and market position. Quality is also identified as having a direct positive effect on profitability (Klein and Leffler, 1981; Schoeffler, Buzzell, and Heany, 1974). The impact of perceived product quality on return on investment (ROI) is positive and significant at above the 95% confidence level in four cases in a study conducted by Jacobson and Aaker (1987) in order to examine consumer durables, consumer nondurables, capital goods, raw and semifinished goods, component and supplies businesses. They identify quality as an endogenous variable which is jointly dependent on other strategic variables and has a consistent and significant positive influence on ROI, market share, and price. Further, price is identified as one of the most common cues of perceived product quality (Scitovsky, 1944). Product quality is identified as a competitive strategy in product differentiation and the attainment of increased market share (Porter, 1980).

Strategy

Social linkages hold some strategic implications for international marketers. Previous research shows that the country of origin stereotype is positively improved for countries with which the potential importer has social linkages (Renwick and Renwick, 1988). Further, importers are most likely to buy from countries with whose nationals they have formed friendships. The current study demonstrates that the perceived product quality is higher for those countries with which there are such social linkages, even after purchasing experience with those countries. The international marketer can use this information to good advantage. In general terms, the country of origin stereotype becomes a matter of consideration with respect to new export markets where the company or brand name are unknown or unfamiliar. In this instance, the country of origin label plays a large role in the purchase decision, and the marketer should use the knowledge of his country's marketing image as a critical part of the marketing opportunity analysis (Darling and Kraft, 1977).

Should the exporter decide to enter a market where his country of origin stereotype is unfavorable, it is suggested that this might be offset by an association with a well-known and esteemed local distributor (Cattin et al., 1982). The social linkage research suggests that an additional variable is important in the marketing opportunity analysis, namely, the percentage of the exporting countries' nationals that are represented in the population of the importing countries under consideration. Any country which is home to a significant percentage of the exporting nation's expatriates should be excluded because it is very probable that the national stereotype has historically been negative (i.e., conflicts typically accompany increased competition with the indigenous population over common resources). However, if a country has a higher percentage of an exporting nation's expatriates but this does not constitute a significant proportion of the population, a positive country of origin stereotype is more likely to be engendered than if a country has a lower percentage of these expatriates. In essence, the number of social linkages forged with these expatriates will be a function of their availability.

A case can be made for foreign postings in markets which international marketers wish to develop. In addition to ability, the need and capacity to make friends might be appropriate criteria for candidate selection. When circumstances necessitate remote contacts (e.g., by telephone, mail, etc.) rather than personal ones, the same individual should serve as the contact with the importer or potential importer. Alternatively, within a given export company, several individuals of the same gender could fulfill this role if they all used the same name when dealing with the importer. These strategies could serve to facilitate social linkages and the associated favorable effect on stereotypes of product attributes.

The findings of this research also suggest that appropriate advertising strategies for international marketers should feature nationals from the exporting and importing nations involved in some culturally relevant activity which implies that they are friends. Beyond the company level, opportunities for increased social linkages can be fostered by foreign trade tours, representation at foreign trade shows, exchange programs, support for the performing arts on foreign tours, and foreign students studying in either country. As the number of social linkages increases the country of origin stereotype is likely to become more positive.

White and Cundiff (1978) indicate that it is advantageous to have a favorable image of product quality if the marketer expects to improve the exportability and marketability of export products. They also suggest that industrial buyers tend to stereotype the quality of products from foreign countries and that this knowledge should form part of the international marketing strategy. Social linkages have been shown to positively influence the perceived product quality even when importers have purchasing experiences with the country in question. As the generalized product level is utilized, it is assumed that the importers respond according to their product purchasing experiences with foreign suppliers (Etzel and Walker, 1974). This suggests a product strategy which would increase penetration into export markets. This involves the selection of export markets which will have a higher ratio of social linkage potential based upon the representation of expatriates in the populations of potential export markets. It also includes the selection of products which can be identified as "value" products, that is, good quality at a good price, since, if the social linkages are there, the products are already stereotyped in a positive direction in terms of the product quality. A converse strategy is also suggested in terms of identifying markets suitable for aggressive pricing and an associated lower level of quality. Analysis of competitive nations' population distribution internationally will serve to identify markets where the marketer is either better represented or more poorly represented and, hence, can more clearly indicate the competitive marketing strategy appropriate to the international market.

Future Directions

Future research into social linkages and country of origin stereotyping would benefit from two modifications to the design used in this research. First, the frequency of purchasing experiences with the various exporting nations should be assessed. Second, a friendship schema which would allow the categorization of the level of social linkage is recommended (La Gaipa, 1977; Renwick, 1985). Future studies could also focus on countries with a higher level of economic development than those examined here.

REFERENCES

Allport, G.W. (1958). *The Nature of Prejudice*. New York: Doubleday Anchor.

Amir, Y. (1969). Contact Hypothesis in Ethnic Relations. *Psychological Bulletin*, Vol. 71, pp. 319-341.

Angelmar, R. & Pras, B. (1984). Product Acceptance by Middlemen in Export Channels. *Journal of Business Research*, Vol. 12, pp. 227-240.

Asch, S.E. (1946). Forming Impressions of Personality. *Journal of Abnormal and Social Psychology*, Vol. 41, pp. 258-290.

Bartlett, F.C. (1932). *Remembering: A Study in Experimental and Social Psychology*. Cambridge: The University Press.

Batsell, R.R. & Wind, Y. (1980). Product Development: Current Methods and Needed Developments. *Journal of the Market Research Society*, Vol. 22, pp. 122-126.

Bilkey, W.J. & Ness, E. (1982). Country of Origin Effects on Product Evaluations. *Journal of International Business Studies*, Spring/Summer, pp. 89-99.

Brewer, M. (1979). The role of ethnocentrism in intergroup conflict. In Austin, W.G. & Worchel (eds.), *The Social Psychology of Intergroup Relations*. Monterey, California: Brooks Cole.

Brigham, J.C. (1971). Ethnic Stereotypes. *Psychological Bulletin*, Vol. 76, pp. 15-38.

Briscoe, M.E., Woodward, H.D. & Shaw, M.E. (1967). Personality Impression Change as a Function of the Favorableness of First Impressions. *Journal of Personality*, Vol. 35, pp. 343-357.

Campbell, D.T. (1967). Stereotypes and the Perception of Group Differences. *American Psychologist*, Vol. 22, pp. 817-829.

Cantor, N. & Mischel, W. (1977). Traits as Prototypes: Effects on Recognition Memory. *Journal of Personality and Social Psychology*, Vol. 35, pp. 38-48.

Cattin, P., Jolibert, A. & Lohnes, C. (1982). A Cross-Cultural Study of "Made in" Concepts. *Journal of International Business Studies*, Winter, pp. 131-141.

Cohen, C.E. (1977). Cognitive Basis of Stereotyping. Paper presented at the 85th annual convention of the American Psychological Association, San Francisco, California.

Darling, J. & Kraft, F. (1977). A Competitive Profile of Products and Associated Marketing Practices of Selected European and NonEuropean Countries. *European Journal of Marketing*, Vol. 11, pp. 519-531.

Etzel, M.J. & Walker, B.J. (1974). Advertising Strategy for Foreign Products. *Journal of Advertising Research*, Vol. 14, pp. 41-44.

Fishbein, M. & Ajzen, I. (1975). *Belief, attitude, intention and behavior: An introduction to theory and research*. Reading, MA: Addison Wesley.

Gaedeke, R. (1973). Consumer Attitudes Toward Products 'Made in' Developing Countries. *Journal of Retailing*, Vol. 49, pp. 13-24.

Garvin, D.A. (1984). Product quality: An Important Strategic Weapon. *Business Horizons*, Vol. 27, pp. 40-43.

Geistfeld, L.V., Sproules, G.B. & Badenhop, S.B. (1977). The Concept and

Measurement of a Hierarchy of Product Characteristics. In Hunt, H.K. (ed.), *Advances in Consumer Research*, Vol. 4, pp. 302-307. Ann Arbor, MI: Association for Consumer Research.

Gilman, S.L. (1985). *Difference and Pathology: Stereotypes of Sexuality, Race, and Madness*. Ithaca, NY: Cornell University Press.

Grace, H.A. (1954). Education and the Reduction of Prejudice. *Education Research Bulletin*, Vol. 33, pp. 169-175.

Grace, H.A. & Neuhas, J.O. (1952). Information and Social Distance as Predictors of Hostility Toward Nations. *Journal of Abnormal and Social Psychology*, Vol. 47, pp. 540-545.

Hakansson, H. & Wootz, B. (1975). Supplier Selection in an International Environment: An Experimental Study. *Journal of Marketing Research*, Vol. 12, pp. 46-51.

Heider, F. (1958). *The Psychology of Interpersonal Relations*. New York: Wiley.

Heslop, L.A., Liefeld, J. & Wall, M. (1987). An Experimental Study of the Impact of Country of Origin Information. In Turner, R. (ed.), *Marketing*, Vol. 8, pp. 179-185. New York: Administrative Sciences Association.

Hill, C.E. & Still, D.E. (1981). Sex Differences in Effects of Social Value Similarity in Same Sex Friendship. *Journal of Personality and Social Psychology*, Vol. 41, pp. 488-502.

Hill, R.M., Alexander, R.S. & Cross, J.S. (1975). *Industrial Marketing* (4th ed.). Homewood, IL: R.D. Irwin.

Hirschman, E.C. (1980). Attributes of Attributes and Layers of Meaning. In Olson, J.C. (ed.), *Advances in consumer research*, Vol. 7, pp. 7-12. Ann Arbor, MI: Association for Consumer Research, 7-12.

Inkeles, A. & Levinson, D.J. (1968). National Character: The Study of Model Personality and the Sociocultural System. In Lindzey, G. & Aronson, E. (eds.), *Handbook of Social Psychology*, Vol. 4, Reading, MA: Addison Wesley.

Jacobson, R. & Aaker, D.A. (1987). The Strategic Role of Product Quality. *Journal of Marketing*, Vol. 51, pp. 31-44.

Jaffe, D.E. & Nebenzahl, I.D. (1984). Alternate Questionnaire Formats for Country Image Studies. *Journal of Marketing Research*. November, pp. 463-471.

Johansson, J.K., Douglas, S.P. & Nonaka, I. (1985). Assessing the Impact of Country of Origin on Product Evaluations: A New Methodological Perspective. *Journal of Marketing Research*, Vol. 22, pp. 388-396.

Jones, R.A. (1982). Perceiving Other People: Stereotyping as a Process of Social Cognition. In Miller, A.G. (ed.), *In the Eye of the Beholder* pp. 41-91. New York: Praeger.

Kaynak, E. & Cavusgil, S.T. (1983). Consumer Attitudes Towards Products of Foreign Origin: Do They Vary Across Product Classes? *International Journal of Advertising*, Vol. 2, pp. 147-157.

Kincaid, W.M. (1970). *A Study of the Perception of Selected Brands of Products as Foreign or American and Attitudes Toward Such Brands*. Unpublished doctoral dissertation, University of Florida, Florida.

Klein, B. & Leffler K. (1981). The Role of Market Forces in Assuring Contractual Performance. *Journal of Political Economy*, Vol. 89, pp. 615-641.

La Gaipa, J.J. (1977). Testing a Multidimensional Approach to Friendship. In Duck, S. & Gilmore, R. (eds.), *Personal Relationships 1: Studying Personal Relationships*. London: Academic Press.

Levine, R.A. & Campbell, D.T. (1972). *Ethnocentrism*. New York: Wiley.

Lillis, C.M. & Narayana C.L. (1974). Analysis of 'Made in,' Product Images: An Exploratory Study. *Journal of International Business Studies*, Spring, pp. 119-127.

Lippmann, W. (1922). *Public Opinion*. New York: Harcourt Brace.

Lott, A.J. & Lott, B.E. (1972). The Power of Liking: Consequences of Interpersonal Attitudes Derived from a Liberalized New Secondary Reinforcement. *Advances in Experimental Social Psychology*, Vol. 6, pp. 109-148.

Mackie, M. (1973). Arriving at Truth by Definition: The Case of Stereotype Inaccuracy. *Social Problems*, Vol. 20, pp. 431-447.

Moreland, R.L. & Zajonc, R.B. (1982). Exposure Effects in Person Perception: Familiarity, Similarity, and Attraction. *Journal of Experimental Social Psychology*, Vol. 18, pp. 395-415.

Nagashima, A. (1970). A Comparison of Japanese and U.S. Attitudes Toward Foreign Products. *Journal of Marketing*, Vol. 34, pp. 68-74.

Nagashima, A. (1977). A Comparative 'Made in' Product Image Survey Among Japanese Businessmen. *Journal of Marketing*, July, pp. 95-100.

Naranya, C. (1981). Aggregate Images of American and Japanese Products: Implications on International Marketing. *Columbia Journal of World Business*, Summer, pp. 31-35.

Nisbett, R.E. & Ross, L.D. (1980). Human Inference: Strategies and Shortcomings of Social Judgements. *Psychological Review*, Vol. 84, pp. 231-259.

Ouchi, W.G. (1981). *Theory Z*. New York: Avon Books.

Peter, J.P. & Olson, J.C. (1987). *Consumer Behavior: Marketing Strategy Perspectives*. Homewood, IL: Irwin.

Phillips, L.W., Chang, D.R. & Buzzell, R.D. (1983). Product Quality, Cost Position, and Business Performance: A Test of Some Key Hypotheses. *Journal of Marketing*, Vol. 47, pp. 26-43.

Porter, M.E. (1980). *Competitive Strategy*. New York: The Free Press.

Rao, C.P., Rosenberg, L.J. & White, R. (1988). Industrial Buyer 'choice criteria' Under Diverse, Industrial Settings. In Bahn, K.D. (ed.), *Developments in Marketing Science: Proceedings of the Academy of Marketing Science Conference,* pp. 132-136. Montreal, Quebec.

Reierson, C. (1966). Are Foreign Products Seen as National Stereotypes? *Journal of Retailing*, Fall, pp. 33-40.

Renwick, F. (1988). Country of Origin Effect: The Development and Field Testing of an Instrument for International Data Collection. Manuscript submitted for publication.

Renwick, F. & Renwick, R.M. (1988). Country of Origin Images: The Effects of Purchasing Experience and Social Linkages Upon Cross Cultural Stereotyping. Manuscript submitted for publication.

Renwick, R.M. (1985). *Competition with Friends: Perceptions, Accounts, and Expectations*. Unpublished doctoral dissertation, University of Lancaster, Lancaster, England.

Rothbart, M., Fulero, S., Jensen, C., Howard, J. & Birrell P. (1978). From Individual to Group Perspectives: Availability Heuristics in Stereotype Formation. *Journal of Experimental Social Psychology*, Vol. 14, pp. 237-255.

Salter, C.A. & Teger, A.I. (1985). Change in Attitudes Toward Other Nations as a Function of the Type of International Contact. *Sociometry*, Vol. 38, pp. 213-222.

Schoeffler, S., Buzzel, R.D. & Heany, D.F. (1974). Impact of Strategic Planning on Profit Performance. *Harvard Business Review*, Vol. 56, pp. 104-114.

Schooler, R. (1971). Bias Phenomena Attendant to the Marketing of Foreign Goods in the U.S. *Journal of International Business Studies*, Spring, pp. 71-80.

Scitovsky, T. (1974). Some Consequences of the Habit of Judging Quality by Price. *Review of Economic Studies*, Vol. 12, pp. 100-105.

Secord, P.F. & Backman, C.W. (1974). *Social Psychology*. New York: McGraw-Hill.

Sell, D.K. (1983). Research on Attitude Change in U.S. Students Who Participate in Foreign Study Experiences: Past Findings and Suggestions for Future Research. *International Journal of Intergroup Relations*. Vol. 7, pp. 131-147.

Tajfel, H. (1969). Cognitive Aspects of Prejudice. *Journal of Social Issues*, Vol. 25, pp. 79-97.

Tajfel, H., Sheikh, A.A. & Gardner, R.C. (1964). Content of Stereotypes and the Inference of Similarity Between Members of Stereotyped Groups. *Acta Psychologica*, Vol. 22, pp. 191-201.

Taylor, S.E., Fiske, S.T., Etcoff, N.L. & Ruderman, A.J. (1978). Categorical and Contextual Bases of Person Memory and Stereotyping. *Journal of Personality and Social Psychology*, Vol. 36, pp. 778-793.

Tversky, A. & Kahnemann, D. (1973). Availability: A Heuristic for Judging Frequency and Probability. *Cognitive Psychology*, Vol. 5, pp. 207-232.

Wang, C. & Lamb, C.W. (1983). The Impact of Selected Environmental Forces upon Consumers' Willingness to Buy Foreign Products. *Journal of the Academy of Marketing Science*, Vol. 11, pp. 71-84.

White, P.D. (1979). Attitudes of U.S. Purchasing Managers Toward Industrial Products Manufactured in Selected European Nations. *Journal of International Business Studies*, Spring/Summer, pp. 81-90.

White, P.D. & Cundiff E.W. (1978). Assessing the Quality of Industrial Products. *Journal of Marketing*, January, pp. 80-86.

Zajonc, R.B. (1968). Attitudinal Effects of Mere Exposure. *Journal of Personality and Social Psychology*, Vol. 9, pp. 1-27.

Chapter 8

Social, Economical, and Political Framework of Joint Venture with Foreign Capital Formation in a Communist System: The Case of the Polish Economy

Krzysztof A. Lis
Henryk Sterniczuk

INTRODUCTION

Communist economies try to join an international division of labor to be involved in an international exchange of technology and to break their own limits concerning innovations, quality of products, and access to capital and markets. The last several years have brought unusual dynamism to that process. From China to Romania, almost every communist country has implemented new regulations. These rules allow an involvement in international business to be easier. Joint ventures, with foreign capital, may be viewed as a new strategy to get a new supply of capital, especially in the form of hard currency and know-how in technology and management. This is possibly the last reserve of communist economy.

After World War II, people had been working hard, relieved from war, with hope for the future. In the fifties an industrialization in most communist countries was paid for by unbelievably poor living conditions in the villages. After Joseph Stalin's death, there was a new hope: communism with a human face and higher standard of living. The seventies meant cheap, easy loans from Western banks, and a dream of developed economy. Simultaneously, the strategic resources have been wasted as always. An output used to be without proportion to input.

The eighties meant disaster in Poland, economic troubles in other communist countries, waking up the Soviet Union, and the lack of chances for more loans. Further collaboration of people with governmental goals became more difficult than ever before. It seems that many strategies have been tried, but an economy system still needs external support in order to survive.

The formation of joint ventures with foreign capital is a new chance, a new strategic resource; it is the way to generate new capital to pay huge debts. The issue is, however, how communist countries are to participate in joint ventures with foreign capital without being exposed to capitalism in their own country. The answer is innovative: let us have joint ventures with foreign capital as so-called "socialized units of economy." This means a venture which is no more than 49% foreign, and no less than 51% state owned, but still communist in its logic.

An examination of the political and economical conditions of joint ventures with foreign capital in a communist system is the main goal of this article. It will show what a long distance communist countries have passed in their business organizations' forms, from a closed economy in the fifties, to joint ventures and limited fully owned foreign enterprises in the eighties. This way one may better understand a current step in an organizational evolution of communist economy.

This paper will present the case of the Polish economy. Poland is the most "capitalist" communist country. Its economy consists of state owned companies as well as private farms, private small enterprises, and foreign owned small companies. In 1987 not less than 75% of agricultural land belonged to private owners and there were about 700 foreign owned companies. There is now a growing interest in state enterprises in joint ventures with foreign capital. There are 20 of them currently. At the same time, normal communist propaganda, power of the communist party, bureaucratic red tape, shortages of everything, lack of hard currency, black market, lack of consequent political solutions, mismanagement and so on, still exist.

Processes going on in Poland show that the ineffectiveness of communist economy is deeply rooted and cannot be easily solved by the implementation of some forms of market systems. This should be taken into account while one evaluates the last reforms in the Soviet Union under M. Gorbachev's *Perestroika*. We may assume that, to a great extent, issues identified in Poland in a position of joint ventures with foreign capital, in the law, and the economy system, may be even more important in other communist countries. Other communist bloc countries have far less experience with market systems. This does not mean that the law regarding joint ventures is the most progressive or market-oriented in Poland; it means, however, that Polish managers, as well as communist party officials, being used to having many market elements in their economy, are possibly better prepared to understand and to deal with market systems, and its social and economical results.

It must be understood that the law regulating joint ventures is the result of internal politics, fighting between different factions which believe in their own "one best way" for the economy. This is why there are so many compromises and contradictions in the law and everyday practice in establishing joint ventures.

Two short cases are provided in an Appendix to show the kinds of economic events discussed, and what the scale of operations is. Both cases present joint ventures in the early stage of formation.

The Polish case may be helpful in an understanding of logic of action in other countries, but cannot serve as a picture for the entire communist-bloc.

EVOLUTION OF POLITICAL CONCEPTS OF FOREIGN PARTICIPATION IN COMMUNIST ECONOMY

Communist countries are now far away from the point where their economy system is governed by market type rules, with a relatively free allocation of domestic and foreign capital. But they are also very far away from the point where they were before Joseph Stalin's death. Intense fights against capitalist forms of economy have been replaced with a compromise type of solution in the last two decades.

The Soviet's communism, excluding the short period of Lenin's New Economic Policy (NEP), used to have a stable, hard-line opinion about capitalist forms of economy. NEP has meant Lenin's flexible approach to an ideology. Economic crisis in the Soviet Union was overcome to a great extent by capitalist organizational forms including foreign direct investment in the Soviet economy.

As economists emphasized, NEP policy have assumed only timely access of capitalist elements to communist economy. "State government have maintained control over rules of a capitalist market and set up rules for a reduction of capitalist forms in communism" (Brus and Pohorille, 1953, p. 83). From the beginning NEP was meant only as a tactic for overcoming internal economic problems, hunger, and extremely poor productivity in newly socialized enterprises. NEP was cancelled as soon as the economic situation became slightly better because NEP's goal was not an affluent society; its goal was, rather, the support of survival of communism.

Later on, in Stalin's period of communist history, the situation was more clear. Market type of economy organizations were one-dimensionally identified with an exploitation and pauperization of the working class. In international relations, British and American imperialism was responsible for exploitation of other countries. Export of capital was, in communist ideology and policy, the worst strategy, leading to the deepest damage in economy systems of other countries. Says Bucharin, "Export of capital is the most convenient method of economic policy of financial groups leading to the easiest way to control of new areas" (Bucharin, 1934, p. 130). Another communist party official expressed the same opinion, twenty years later; making comments on international policy, G. Malenkow said: "Any United Kingdom's enemy did not cause its so deep damage and did not remove this country from its Empire a piece after piece as did its American friend" [through export of capital] (Malenkow, 1952 p. 22).

Obviously this kind of opinion of the Soviet communist leaders affected an economic policy of new communist countries, established under the Soviets' supervision, after World War II. Poland was one among them. Before 1939, Polish industries had shared domestic-foreign ownership. For example, mining and steel industries belonged to foreign capital in 52.5%; oil industry in 87.5%; electric industry in 66.1%, and chemical industry in 59.9% (Minc, 1946, pp. 14,15). That situation was perceived by the new communist government as a serious threat to a communist order, as well as to national independence (Minc, 1946, p. 14). Immediately after the war, economic policy of the new Polish government did not exclude an economic collaboration with western countries. The policy conducted in that period reflected a coalition type of government: social democracy and communist parties. President of the Central Planning Office, a leading agency for economic policy design, Cz. Bobrowski, expressed the socialist position this way: ". . . Foreign loans are important conditions of a realization of our plans. We can not think about our goals achievement without a participation of foreign capital (. . .) There is no reason not to assume that a process of slow, but constant inflow of foreign capital will continue" (Bobrowski, 1946, p. 42). In that time, even the communist faction assumed that about 20% of capital necessary for an investment would come in the form of western loans (Minc, 1946, p. 28). This moderate policy of the coalition government dramatically turned around in 1948 after a liquidation of the Central Planning Office, a unification of political parties, and a total stabilization of a communist order in Poland. In the next period of hardline communist policy, only limited trade between western and eastern countries was permitted, while other forms of international economic relations were excluded.

A new period began after Stalin brought the international collaboration again to public debate. The inability to obtain hard currency became among one of the most difficult obstacles to overcome for economic development. Despite that, direct investment of foreign capital in the Polish economy was still not accepted, due to the strong dogma of communism. In order to cope with shortages of many products and technical stagnation, the Polish and other communist governments decided to buy licenses rather than to give foreign companies access to domestic economies. In the sixties, and with intensity in the seventies, we observe a large purchasing of patents, licenses, machines, and whole technological lines for manufacturing. Purchases were made mostly as a result of large loans from western countries. Architects of that policy believed that they would be able to pay back the loans through the export of goods and services from modernized industries. Nothing like that happened. The investments never paid back in appropriate proportions. It was an expensive solution, and never completely successful. Causes of that are very deeply rooted in the politically, instead of economically, driven mechanisms of an allocation of capital in a communist system, in lack of market orientation of enterprises, inconvertibility of currency, and low effectiveness of capital investment. This way, Poland became economically bankrupt in the seventies. Loans became very hard to get, and goods did not have an access to western markets because of their quality and economic troubles of Western Europe in that time.

Economy crisis and social tensions demanded new solutions. Again, to a certain extent, as in NEP-time, a new strategy was desperately needed in order to survive. Social pressure on the betterment of life conditions, lack of easy reserves, and foreign debts contributed to crises that, after several years, brought more flexible policy to many areas of economy and social life.

CHANGES OF LEGAL REGULATIONS REGARDING FOREIGN CAPITAL FUNCTIONING IN POLAND

Almost all communist countries have permitted foreign capital for a direct investment on their territories. Yugoslavia started in 1967, Romania joined in 1971, then came Hungary in 1972, Poland in 1976, China in 1979, Bulgaria in 1980, more recently, Czechoslovakia in 1986, and a year ago (January 13, 1987), the Soviet Union. The dates show how relatively new this phenomenon is in international economic relations.

Advocates of new policy have emphasized that joint activity with foreign investors at the territory of communist countries provides an opportunity for the concentration of factors of development such as capital, productive capacities, technology, and the managerial expertise. This way, partners make possible the completion of projects far beyond their individual potential (Burzynski, 1987).

There are certain common features in the joint venture legislation of the communist countries; however, there are also essential differences (Buzescu, 1984; Scrirven, 1982): Polish legislation of 1976 differed from other communist countries by allowing one hundred percent foreign-owned investment as well as by mixed companies (Burzynski, 1987).

Looking at the legislation process historically, we see three regulations appear at the same time:

- the Decree of the Council of Ministers of May 14, 1976 concerning the issuance of permits to foreign legal and natural persons for conducting various types of economic activity (Monitor Polski, No. 25, Item 109)
- the Decree of the Ministry of Finance of May 26, 1976 permitting the opening and use of bank accounts for those who conduct economic activity in Poland (Monitor Polski No. 15, Item 110)
- the Decree of the Minister of Finance of May 26, 1976 concerning permits for foreign exchange operations by mixed capital companies

This set of new rules opened a potential access of foreign capital to the Polish economy; however, it was only a limited opening. Several areas were designated: crafts, retail trade and catering, hotels, and other minor services. Later on, in 1978, it was spread to all manufacturing and service activities. According to an amendment issued on December 1, 1978 (13 Journal of Laws, Item 138), a permit for a foreign enterprise may be granted for a period longer than ten years, if justified by the nature of the economic activity, the value of the investment, and the projected investment return. Local public administration was in charge of issuing permits. Prior to obtaining a permit, the foreign investor had to present a cost projection for an investment, a commitment to cover the full cost of the investment in convertible currency, and a bank certificate proving that thirty percent of the estimated investment cost had been deposited in convertible currency.

The Decree of the Minister of Finance is interesting for us. It provides an opportunity for mixed capital companies in Poland. According to its provisions, foreign legal persons or associations of

Poles living abroad could enter into an agreement with the Polish state enterprises, cooperatives, or non-profit organizations conducting business activities in the form of a mixed capital company. In this case, the Minister of Foreign Trade is supposed to work in conjunction with the Minister of Finance, and the Minister responsible for the given branch of national economy was in charge of issuing appropriate permits. A necessary capital contribution of the foreign partner had to be made in Polish currency, obtained from a documented exchange of convertible currency. Non-pecuniary contributions were also allowed, but for no more than 50% of the total contribution. In both cases (wholly owned and mixed), foreign partners could receive part of their profits, derived from the operations, in convertible currency. At a particular tax year, only 9% of the value of the capital contributed by the foreign partner was allowed to be converted into hard currency. This rule did not apply when 50% or more profit was generated by export goods or services.

The Decree of 1976 made a good beginning to the next important steps, but did not accomplish any significant achievements in itself. Both sides, foreign investors as well as the governmental agencies and local authorities, had not been prepared to understand the complicated issue of a communist system from one side and rules of market from the other side. Fortunately, the disappointed Polish government has not cancelled the law, but has replaced it with a new one which was more liberal and progressive, at least in its intentions.

The next step was five major decrees and resolutions established in the spring of 1979:

- resolution of the Council of Ministers of February 7, 1979 on Establishing Business Enterprises with Foreign Capital Participation in Poland, and their operations (Monitor Polski, Item 36, No. 4, 1979)
- Decree of the Minister of Finance, dated March 28, 1979 regarding the Permission to Open and Maintain Bank Accounts by Foreign Corporations and Individuals Conducting Business Activities in Poland. (Monitor Polski, Item 67, No. 10, 1979)
- Decree of the Minister of Foreign Trade and Maritime Economy of March 28, 1979, on the Permission to Conduct Certain

Foreign Trade Activities by Foreign Corporate Bodies and Individuals (Monitor Polski, Item 68, No. 10)
- Decree of the Minister of Labor, Wages and Social Benefits of May 30, 1979 regarding Some Principles of Employment and Compensation of the Employees of Limited Liability Companies with Foreign Participation (Monitor Polski, Item 88. No. 15, 1979)
- Decree of the Minister of Finance of June 18, 1979 Concerning Financial Operations of Business Enterprises with Foreign Capital Participation and on Permits for Conducting Certain Foreign Exchange Operations (Monitor Polski, Item 97, No. 16, 1979)

As one of the most outstanding advocates for joint ventures, Dr. A. Burzyniski emphasized the fact that a main act (Resolution on Establishing Business Enterprises with Foreign Capital Participation in Poland and their Operations) was issued by the Council of Ministers, not by the Minister of Finance as it was before, is the most important feature of the new set of regulations. The Council of Ministers is the highest authority in the government of Poland. It means that a resolution was issued by the whole government, and that other branches and functional Ministers had to follow this resolution. We have to remember that coordination is extremely difficult in this government where particular Ministers govern very separated, limited areas of the economy system and public matters. Every office cares only for its own area and fights with the others for the limited resources centrally distributed: how much money, how much hard currency, how many employees, what kinds and sizes of new investments and so on; everything is short. If one gets more, another will have less. It was a zero-sum game. If a decree was issued by the Minister of Finance, the message was that his office was interested in an increase of export of goods, and in a generation of hard currency. Regulations, issued by one Minister, but regarding the entire economy system, used to be hardly followed by others when they were not covered by strong sanctions.

This must be strongly emphasized in order to allow Western readers to understand how significant it was that regulations had been issued by the Council of Ministers. The new set of rules sent the message to foreign investors that the government was willing to progress in that issue. In terms of easiness or convenience for foreign investors, however, the new regulations did not bring many significant changes, and in fact, required even more detailed procedures for permit applications.

The application was to be submitted by the managing director of a state enterprise or by the President of the cooperative. A special statement was required from the relevant Polish foreign trade agency, concerning the prospects for future export of the joint venture products to convertible currency markets. The new regulations limited the scope of activities of the prospective joint venture to that allowed to cooperatives and small industrial organizations being under the authority of a regional public administration. That means that only small and medium scale organizations, functioning outside of central priorities, might be involved in joint ventures with foreign capital. The permits for remission of convertible currency were to be issued only within the limits of the convertible currency reserves of the joint venture.

The regulations of 1979 shared their limited effectiveness with acts of 1976. They had very limited impact on the Polish economy. A number of wholly-owned small foreign businesses were established in that time but because of their separation from the main stream of the Polish economy, they did not play any significant role. Despite this, they have been followed by a dogmatic discussion regarding social justice, wages, and equity, and other ideological issues.

The next step in the regulation was made in 1982. This time it was the law established by the Parliament of Poland.

- Law on Banking of February 16, 1982 (Dziennik Ustaw No. 7, Item 56, 1982)
- Law on Foreign Corporate Bodies and Natural Persons' Economic Activity in Poland (Dziennik Ustaw, No. 7, Item 56)

These laws again did not bring significant novelties in their content, but the fact that they had been issued by the Parliament changed the position of the issue in the economy system as well as in political meanings. Now, Poland has regulation acts that are supposed to be followed by enterprises as well as by the government and its agencies. It is the highest level regulation that is possible in a business law. As the result of these rules, a majority of 695 foreign owned companies, currently working, were established. The topic of wholly owned foreign small businesses in Poland is interesting in itself, but the focus here is on joint ventures.

The last step in the evolution portrayed here was undertaken by the Parliament in 1986.

– Law on Companies with Foreign Capital Participation of April 12, 1986 (Dziennik Ustaw No. 17, Item 88)

This time it was a specific joint venture law. Thus, a ten year period of evolution of regulation regarding foreign involvement in the Polish economy was closed.

Characteristics of some issues emerging from new laws and regarding a position of joint ventures in political and economy systems will be the subject of the next section of this chapter.

JOINT VENTURE WITH FOREIGN CAPITAL AND ITS POSITION IN THE POLISH ECONOMY

Joint ventures with foreign capital are really a new phenomena in the Polish economy. Currently, more joint ventures are in the process of leading to registration than functioning. According to the Polish Chamber of Foreign Trade, about two hundred Polish state-enterprises are looking for an opportunity to perform business together with some Western partners. About one hundred negotiations are being conducted. A large diversity of Polish industries are involved in this process. If this rapid process continues, in the near future we may expect many joint ventures and a new stage in East-West collaborations may be opened. So far, it is difficult to deliver spectacular examples, showing how this new form of conducting business is working, in an environment of many "red tape" obstacles, centralization of management in many areas, and political interventions of the communist party.

The Polish government is willing to attract foreign capital, due to billions of foreign debts. But, to be willing does not yet mean to have work done. In this case, in the current law and in the role of the government offices in the registration process, many obstacles are hidden. They are a result of an ambiguity which joint venture with foreign capital brings to an ideological order of the communism, and by some kind of "learned ineffectiveness" of the communist system in dealing with foreign capital and market economy.

At least two contradictions regarding an essence of a joint venture as a business are possible to note now before some other problems appear.

Problem of Equity of a Joint Venture with Foreign Capital to Other Fully State Owned Businesses

According to article 26, Item 1, joint ventures with foreign capital participate in the economy the same way other businesses, state owned or cooperatives do. For state enterprises there is a special term in Poland: "socialized unit of economy." A joint venture is treated as a "socialized unit of economy." In relationships with other organizations of Polish economy and with the governmental agencies, joint ventures are regulated through the law and orders or resolutions that every business is obliged to follow, if the Law on Companies with Foreign Capital Participation does not determine it in another way. This type of positioning of joint ventures in Polish economy creates many negative consequences. Polish (Rajski, 1986), as well as foreign researchers (Juregensmeyer, 1987), criticize this. A difficult economic situation in Poland, a continued centralized management system, and repeated changes in regulations regarding the economy create a climate that does not encourage foreigners to invest and form larger scale businesses. Article 32, Item 2 says that the Polish Law regulates the compensation of employees, insurance and other issues. The rule itself is quite obvious, but a specific regulation that follows it will probably contribute to many management problems of joint ventures. There are several taxes that joint ventures have to follow as do other socialized units of economy. But again, the issue is not the tax itself but, rather, in its repressive function in a communist economy.

Let us take as an example employees' compensation. Every "socialized unit of economy" has to pay an additional special tax when it increases employees' wages above a certain level connected with the previous year and an increase of its productivity. For example, in the market system, an entrepreneur is free to increase wages of employees due to their higher productivity, or higher prices on the marketplace for his goods, or for any other reason. In the Polish communist economy of 1987, the following situation was possible: an increase of wages above 12% of the 1986 level would cost an enterprise a 500% tax of increased amount of money. But, it does not mean that every enterprise might have had to pay 500% in such a situation. There were hundreds of exceptions caused by five different rules regulating this issue, and by individual bargaining of managers with the state agencies. This additional tax on wages was designed to prevent a fast development of wages that would contribute to greater inflation, lowering company profit, and also lowering the governmental income because of lower taxes on profit. Another goal was, possibly, some kind of economic education of enterprise. Wages are a cost factor; if you have enough money you may pay more. It would be quite easy to increase company cash inflow just by increasing prices of goods, due to very deep imbalances at the market. An increase in wages is very expensive now in enterprises because of this additional tax. For many reasons, the mechanism almost always does not work at all. Wages, as well as prices of goods, are growing out of proportion. This is simply because of a lack of consequent policy issued by the government. Wages have always been a political issue in communism. The increase of wages in order to create the impression of betterment was long practiced as the government's strategy for keeping the social peace for as long as possible. There are still some elements of that strategy in government actions. Summing that up, the tax was implemented in order to develop an effectiveness of the economy at the micro level. But very likely this creates severe results. The issue is that, because of heavy taxation, an enterprise has actually very limited investment funds even while working very successfully. Joint ventures are in the same boat but are typical market-oriented businesses as well, at least in the assumptions of Western partners. The repressive type of taxes mentioned above prevents free allocation of resources, inside an enterprise, according to the best possible profit that they would generate. There is no possibility in this system to minimize wages cost per unit of production without an additional tax punishment. Without tax punishment, an increase of individual wages is almost impossible, but with it, development is hardly possible. Business development is extremely expensive. Business goals and rules governing communist economy seem to be in conflict again.

Found of Foreign Debts is another difficult problem. The tax is established as 2% of value of production means counted at the end of the previous year. This tax is supposed to be withdrawn from funds saved for business development. It is a controversial problem whether or not a joint venture which did not contribute to the development of Polish debts should decrease its development opportunities in order to reduce debts. There is no question that debts are affecting the economy system every day, but money for business development is supposed to take priority, otherwise, all incentive for involvement in the economy in international business will soon die. The reverse approach, typical for conservative policy, would be more understandable in the Western business world, with lower taxes, faster development, higher corporate income, and higher government income as the result of it.

Another controversial issue concerns an income tax. According to Article 30, Item 1, this tax is paid from a verified profit made in the previous year, increased by costs and waste treated as unacceptable by orders of Polish Council of Ministers. What does "unacceptable costs" mean? The definition of an unacceptable cost differs according to the order of the Council of Ministers. The idea of an unacceptable cost was implemented in order to prevent inefficient practice in enterprises. When a company's profit is not the desired value, there is no competition, or demands are higher than supply on the market, it is quite easy to lose an economic approach to business. To remain a necessity for savings of production and other costs, the government established the factor of unacceptable costs. Reaction of managers to it is in the form of a kind of game. They have learned how to call particular costs to avoid punishment. For business people operating in market type conditions, this idea and its implementation through changeable rules issued by the Council of Ministers looks very strange. It creates an insecure feeling, as profit is unknown until its revision, or profits may suffer from different interpretation by managers and the government as acceptable or unacceptable costs.

Joint ventures treated as "socialized units of economy" have to pay typical taxes, some examples of which were mentioned above.

Some of these taxes play the role of factors increasing prices in order to limit demands. Many taxes are collected by the government agencies in order to help the economy develop well-balanced areas of the market. So far, those areas do not exist, but prices of many goods are high, above any proportion of production costs or reasonable profit. Higher prices of many goods do not contribute to higher corporate profit. Sometimes more than 40% of the price value is generated by the taxes. Profit obtained in this manner flows to the state. In consequence, however, this practice lowered demand to a certain extent, which may affect a profitable scale of production of particular goods. It is a significant issue for a joint venture strategy design, and again, an absolutely strange idea for managers trained in market economy.

One Dimensional Pro-Export Focus of the Law

The second set of problems concerns very strong pressure on export of goods by a joint venture with foreign capital. Export is a preferred goal of joint ventures and is an exclusive source of hard currencies. There are several solutions that are concentrated on a pro-export motivation.

In Poland, the Minister of Foreign Trade is in charge of permit issuance for organizing joint ventures with foreign capital. A joint venture, in the governmental assumptions, is to be a vehicle for hard currencies gain. As J. Rayski emphasizes "This type of presumption will probably result in lowering of interests in investing in Poland because for many potential investors full access to the Polish domestic market may make sense" (Rajski, 1986, p. 18). Motivation for export, built into the law system concerning joint ventures, is enormous in the Polish case. For example, profit of joint ventures is taxed at a rate of 50% (that is, 15% lower than in other socialized units of economy), but each value of export in relation to entire values of sold goods and services results in a lowering of 0.4% tax. For example if a company exports everything that it produces, the tax will be lowered until 10%. This is a significant difference. Another strong incentive for export is included in tax of gross sales; goods and services exported are free from this type of tax. The result of it may appear in the form of coerced export which may be in conflict with national economy goals. Again, the issue seems to be that the government is not giving an opportunity to make business more flexible for domestic and foreign companies to make profit. Instead, the government wants to have an opportunity to gain hard currencies. Some other communist countries were able to avoid this sort of greedy approach. For instance, they allow joint ventures to transfer part of the profit in hard currencies independent of the kind of money the company earns on the domestic market. Sometimes the company is allowed to buy and export, free of custom duties, goods equal to a profit gained on the domestic market.

Generally speaking, the joint venture problem in Poland, regulated by the law of 1986 and earlier provisions, is inconsistent between joint venture as a vehicle for hard currency and joint venture as business investment for maximization of profit over a long-time period. Existing rules prevent long term commitment. The limited period of time that a company is permitted to operate, and the limited rights for selling of ownership to another person or company, even to family members of the current owner reinforce that very strongly. Currently, any change in ownership requires prior permits from the Minister of Finance, Minister of Foreign Trade and other institutions.

Current regulations hardly help a company to do successful business which would contribute to lower hard currency income that the government might obtain.

CONCLUSION

The solutions undertaken in the last Polish regulations on joint ventures with foreign capital present an enormous struggle for flexibility and dynamism in the economy system. What can be observed currently is a set of inconsequential and temporary solutions that probably reflect a crisis of thinking about the limits of a communist economy and its identity, as well as about market economy itself. Misunderstanding of business lies at the base of the regulations. It seems that in official perception, business is still something dirty that takes advantages through the fact of its existence itself. No matter how high taxes are raised, and how many limitations will be set up, business will make money and will survive. The Polish government has training in this matter while dealing with its own class of capitalists. The private sector in the Polish economy has always been the strongest among communist countries. Despite taxes acting twenty years back to reduce that group, entrepreneurs were able to build larger houses and buy more expensive cars. But while they used to sell on the domestic market, where demands have always been greater than supply, they prospered on the foundation of the black market and bribes, as well as enormously hard work. Foreign capital, in terms of a direct investment of foreign corporations is not the same case. It needs a clear chance for a good future, reasonable risk, and economic reality that may be at least partially predicted. Current regulations hardly offer those simple things.

Joint ventures with foreign capital have been incorporated into the communist system as "socialized units of economy" and this way have become equal to any other organizational form regulated by the government. Probably this way the greedy face of capitalism will not appear, but it is also highly possible that great innovation, flexibility, and advantages coming from competition will not appear either.

The current approach creates monopolistic units through a system of issuance of permits, which take a long time and are granted according to the definition of economy needs made by the governmental officials, but not by the market process. Newcomers are incorporated into the old structure and are governed by old rules. This is well known, in the professional environment of economist and organizational researchers, that most of the ineffectiveness of communist economy comes from institutional rules and forms of the governmental regulations. Joint ventures meant hope for modern effective enterprises and export of modernized high quality goods. They are under the control of the same rules that maintain ineffective enterprises and the entire economy.

How should effectiveness come to the economy? Automatically, together with several millions of dollars and foreign names? This is the next misunderstanding of the issue. Joint ventures may be some kind of vehicle for mediocre development of hard currency inflow. In their current position in the economy, they do not have a chance to bring new patterns of management and hard priorities for efficiency and quality to the economy. The way joint ventures are incorporated into communist economy seems to show that they are perceived by some strong factions of the government as a replacement for loan-strategy conducted in the seventies. Hard currency is welcomed, export is welcomed, but development and freedom for capital allocation is under the control of the government. The old paradigm of communism, emphasizing that the central government is a better allocator of capital than market mechanisms, is still in charge.

It is very possible that the law of 1986 will have a very short life and will be replaced by a new one. Currently a process of reviewing the new legislation is going on. Again, it is highly possible that coming regulations will offer more flexibility and better conditions for real business.

Progress is slow but constant. As shown by the situation present in Poland, forces for change are already inside the communist economy.

REFERENCES

Bobrowski, Cz. (1946). *Przmowienie na XI Sesji KRN.*

Brus, W., & Pohorille, M. (1953). *Zagadnienia Budowy Ekonomicznych Podstaw Socializmu*, Warszawa.

Brus, W., & Pohorille, M. (1953). *Okres Przejsciowy od Kapitalizmu do Socializmu*, Warszawa.

Brzezinski, B., & Gluchowski, J. (1987). "Regulaeje Finansowe Wlnowym Prawie o Spolkach z Udzialem Zagranicznym." *Panstwo i Prawo.*

Bucharin, N. (1934). *Imperialism a Gospodarka Siwiatowa, Warszawa.*

Burkhardt, A., (1982). "Promotion of East-West Joint Ventures by Governments of Western Countries." *The International Business Lawyer.*

Burzynski, A., (1987). Evolution of Polish Investment Law. Paper presented at Forum for Promotion of Joint Ventures, UNIDO, Warsaw, Poland, October 12-15.

Buzescu, P. (1984). "Joint Ventures in Eastern Europe." *American Journal of Comparative Law.*

Juergensmeyer, J.C. (1987). Perspective on the Prospect for Joint Ventures as a New Form of Cooperation Between East and West. Paper presented at Forum for Promotion of Joint Ventures, UNIDO, Warsaw, Poland, October 12-15.

Kozinski, J. (1987). Joint Venture in Poland, Case study formula. Paper presented at Forum for Promotion of Joint Ventures, UNIDO, Warsaw, Poland, October 12-15.

Lange, O. (1957). *O Niektorych Zagadnieniach Polskiej Drogi do Socjalizmu.* Warszawa.

Lis, K. (1987). Joint Venture – Management Problems. Paper presented at Forum for Promotion of Joint Ventures, UNIDO, Warsaw, Poland, October 12-15.

Malecki, N. (1987). The Conditions of Foreign Capital Functioning in Socialist Countries (A comparative study). Paper presented at Forum for Promotion of Joint Ventures UNIDO, Warsaw, Poland, October 12-15.

Malenkow, G. *Referat na XIX zjazd KC WKP(b)*.

Minc, H. (1946). *O Przejeciu na Wlasnosc Panstwa Podstawowych Galezi Gospodarki Narodowej*. Lodz.

Rajski, J. (1986). "Nowe Prawo o Spolkach z Udzialem Zagranicznym." *Parntwo: i Prawo*.

Scriven, J. (1982). "Co-operation in East-West Trade: The Equity Joint Venture." *The International Business Lawyer*.

Stalin, J. (1952). *Ekonomiczne Problemy Socjalizmu w ZSRR*. Warszawa.

APPENDIX:
A comparison of two joint venture projects*

	LIM Joint Venture Ltd.	Technodiament
1. Partners	1) LOT Polish Airlines, a Polish State Enterprise 2) JLBAU GMbH, an Austrian limited-liability company 3) MARRIOTT International Hotels Inc., an American Corporation	1) PTUPK TECHNOKABEL, a Polish State Enterprise 2) CIE JMPEXMETAL, a Polish foreign trade agency 3) FLT et METAUX, a Belgian joint stock company
2. Business	– completing & operating the LOT Passenger Service Center in Warsaw, Poland – providing the Center's service to the local and export markets	– manufacture, regeneration and marketing of wire drawing dies and tools made of natural and artificial diamonds and ultra-hard materials
3. Duration	27 years after the date of its registration	25 years after the date of its registration
4. Equity Participation	The share capital – 15 million of Austrian Shillings, that means 100 indivisible shares with a par	The company's initial capital amounts to 28 million of Polish Zloty (ZL) and it is divided into

*This appendix was prepared on the base: J. Kozinski, Joint Venture in Poland, Case Study Formula. Paper presented: Forum for Promotion of Joint Ventures, UNIDO, Warsaw, Poland, October 12-15, 1987.

	LIM Joint Venture Ltd.	Technodiament
	value of AS 150 thousand per share; the equity ratio: Polish partners – 52; Austrian – 24; American 24 percent	40 shares with a par value of ZL 700 thousand per share; the equity ratio: Polish Partners: Technokabel – 52.5; JMPEXMETAL – 2.5; Belgian partners – 45 percent
5. Project Financing	In addition to the above the joint venture shall: – secure the external finance in the form of convertible currency loan by Girozentrale Bank in Vienne. The amount of the loan will be, AS billion and its terms are to be specified in a separate credit agreement between LIM Joint Venture Ltd. and the Austrian bank. – secure the subordinated loans from shareholders amounting to AS 135 million	In addition to the above the joint venture will take up two separate credits: – ZL 25 million from the Polish bank to buy selected machines in Poland; – US $370 thousand from a foreign bank to import machines.
6. Management	The authorities of LIM Joint Venture Ltd. are: The General Assembly of Shareholders: the highest decision-making body; The Supervisory Board: five members appointed or elected for a term of two years; The Board of Management two members: Director General designated by LOT, and a Financial Director designated by MARRIOTT	The authorities of Technodiament are: The General Assembly of Shareholders: the highest decision-making body of the joint venture, its resolutions require a simple majority for approval; The Supervisory Board shall consist of three members appointed for a year-term: one by Technokabel, and one by FLT et METAUX, and one to be elected by the company's employees; The Board of Management shall consist of three members: a Managing

	LIM Joint Venture Ltd.	Technodiament
		Director designated by Technokabel, a Financial Director designated by FLT et METAUX, and a Technical Directory Designated by Technokabel
7. Labor Requirements	At the first five years the total employment will be around 850; supervisory personnel, 60; the company will employ up to 30 foreign citizens in year 1989-1992 with substantial reduction after 1992.	In the second year of its operation when the company reaches the feasible normal capacity the total employment will be 26. The company does not intend to employ foreign citizens.

Chapter 9

The Political Element and Marketing in Grenada

Lionel A. Mitchell

INTRODUCTION

Marketing is relevant to all nations, large and small, developed and developing, capitalist and socialist. At the macro level, marketing is defined as planning, producing and delivering goods and services and a standard of living. It looks at the economy's entire marketing system to see how it operates and how efficient and fair it is. Micro marketing, on the other hand, examines individual firms within the economic system to see how they operate or how they should function (McCarthy and Shapiro, 1975, pp. 4-5). The emphasis differs between the two. Nonetheless, marketing has domestic and global appeal. It also plays a critical role in economic growth. It makes possible economic integration and full utilization of the productive capacity that developing nations possess (Drucker 1971, p. 4).

Bartels (1968) noted that marketing has universal technology and Walvoord (1983) confirmed that foreign trade is similar to domestic expansion, but more intricate. The majority of nations, if not all, engage in both domestic and international trade but the mixture of the two varies from one country to another according to stage of economic development and other factors. The reasons for trade, whether domestic or international, include, of course, the satisfaction of wants, the fulfillment of objectives, the improvement in the standard of living and the effective and efficient utilization of individuals' and countries' resources.

International trade accounts for a large part of the world's gross national product and is one of the fastest growing areas of economic activity (Terpstra, 1983, p. 25). This motivates the desire of most countries, even small developing ones that are resource based, to get involved in international marketing as a means of growth and survival. Export marketing is a means to an end for a great number of countries, particularly developing ones. This had led countries like Grenada under its assassinated Prime Minister Maurice Bishop to distribute their products wherever possible. Hence Grenada's association with the U.S.S.R. was a very recent development connected with its objective of rising out of the poor, underdeveloped country category and to improve the standard of living for all of its people. The United States' intervention in 1983 and the subsequent parliamentary election in 1984 changed the emphasis on the elements of the environment for marketing in Grenada. Thus, this environment must be examined in order to understand marketing in Grenada.

There are many ways of classifying the environment of marketing. One classification is in terms of controllable and uncontrollable variables. The controllable elements refer to the marketing mix while the uncontrollable variables fall into the following categories:

1. cultural and social environment
2. political and legal environment
3. economic environment
4. existing business structure
5. Resources and objectives of the organization (McCarthy and Shapiro, 1975, p. 88).

These variables are increasingly important in our dynamic national and world environment. Decision makers' actions may affect some or all of them. Nevertheless, the emphasis of this study will be on the political, especially government actions, in the context of the relationship with the business structure, cultural, economic and social factors.

The role of government in marketing covers three areas. They are: regulatory, facilitative or promotional, and entrepreneurial. The regulatory aspect seeks to establish and to maintain an orderly framework for economic activities and growth, and the regulation of economic resources in the "public interest." The facilitative or promotional aspect entails more direct involvement and possibly costs more. Organizations are established to provide basic services, infrastructure, and other assistance needed by private enterprise. The entrepreneurial aspect entails the most direct involvement and could require the greatest outlay. This involves active participation in ownership and operation of companies. There are certain goods and services that would not be provided were the government not to do so, because at a certain point in time they may not be in sufficient demand, or the costs for producing them may be perceived by private entrepreneurs as too high or the returns too low.

There are differences in the type, degree and nature of the regulatory role from one country to another. Moreover, many people believe that, among other things, the paperwork involved in government regulation is especially harmful to small business (Peterson and Peterson, 1981). This would be the overwhelming size of business to be found in Grenada and other developing countries. Further, Levi and Dexter (1983) suggest that government involvement is an obstacle to the economy and thus is contrary to the presumed "public interest."

The facilitative or promotional role has often been criticized, sometimes even by those whom it was meant to help. It is criticized, too, by those who do not get the benefit but may be competing with those that do. This is especially the case in Grenada and other developing countries where government funds are very limited and priorities have to be established. There is a tendency for these priorities to be perceived in terms of partisan politics.

The entrepreneurial role of government varies from one country to another and from one period to another becoming larger or smaller depending upon stage of development and other circumstances. Nevertheless, Bennett (1967) suggests that this role needs consideration.

It should be fairly clear now that the regulatory, facilitative and entrepreneurial roles of governments differ according to a country's stage of development, size, objectives, needs, resources, competencies and economic and political ideology. Hettich (1983) has demonstrated that governments' decision making and policies do not necessarily correspond with economic cost-benefit analysis. Thus, Wells, Jr. (1977) has suggested that it is of prime importance to understand the *aims* of a government as this could have a great effect on negotiations.

OBJECTIVES AND METHODOLOGY OF THE STUDY

The objectives of this study are an assessment and evaluation of the importance and magnitude of the political element in Grenada's marketing and the achievement of its mission and goals. It compares the political element in the marketing framework under different governments that espoused different ideologies. It examines the role of government in the development of Grenada, a very small independent country, and draws lessons from Grenada's experience that might be useful and applicable to other small countries.

Given the objectives of the study, the methodology is based on library research, case studies, interviews and discussions with government officials, executives of government and semi-government bodies, managers and other business people. It is descriptive and analytical and it encompasses principles of marketing within a theoretical framework.

ASSESSMENT AND EVALUATION

The environment of marketing consists of the internal (to the organization or country) and the external (to the organization or country) factors. The elements of the environment are cultural, competitive, economic, legal, political, social and technological. These elements and their strengths/weaknesses play an important role in macro marketing practices and effectiveness. For example, the legal and political factors weighed strongly in the United States Administration's hostile attitude and negative decisions toward Grenada during the period 1979-1983. The U.S.A. did not accept the legality of Grenada's government under Maurice Bishop without a legitimization of it by a general election. This was exacerbated by the left leaning or socialist politics of Maurice Bishop's New Jewel Movement Party. On the other hand, the United States intervened in Grenada in October 1983. It quickly pushed for an election in 1984 which rectified the legal and political weaknesses from its perspective.

Political and Economic Factors and Development

The political and economic factors as determinants of development are tied very significantly to a global marketing communication process. Development has been described as a process that establishes a balance among three goals: control over nature, over national destiny and over one's self. National development is, therefore, the process of providing the means and opportunity for improving the living standards and the sense of equality of all the inhabitants. It is not a simple economic process. It is a blending of the economic, social and political processes within a specific framework and context. Grenada is an open economy that is dependent on trade with industrialized countries. Whatever happens in and with these countries has an immediate and severe impact on Grenada's economy. Therefore, as part of its marketing communication Grenada must engage in international marketing campaigns to put across important points about itself to the citizens of other countries. Many of the principles and concepts of micro marketing are applicable in macro marketing. These include mission, objectives, planning and strategy, implementation, control, feedback and review.

Politics and Marketing Prior to the U.S. Intervention

Grenada's mission during the period 1979-1983, set by former Prime Minister Maurice Bishop, was to break out of its historically determined role as exporter of cheap agricultural products to a handful of industrialized countries at terms set by the purchasers. The goals were to develop economic self-reliance through the production of local foods and import substitution; to diversify and to expand production; to embark upon social development programs that would eliminate or alleviate poverty, inequality in income distribution, unemployment and economic and social injustices. Thus, the mission and goals were clearly pro-socialist in spirit and intent.

Nonetheless, Bishop had realized the problems and limitations as well as the importance of marketing. In 1981, he acknowledged: "The State Sector alone cannot develop the economy, given the very low level of technology available, the limited human resources, the lack of capital, the lack of marketing expertise, the lack of promotional capacity" (Epica Task Force, 1982, p. 75).

Bishop believed also that the government was the chief agent in attempts to break down the wall between urban and rural life and in assisting in the creation of national markets. The modernization of rural life requires: more efficient distribution, changes in land tenure in the regions, changes in certain cases in government agricultural price policies, expansion and improvement of agricultural credit, build-up of institutions such as producers' cooperatives and food processing firms, and programs of popular cooperation and community development (Rostow, 1971).

Grenada government under Maurice Bishop had embarked upon all of them with the exception of the expansion and improvement of agricultural credit and more efficient distribution. The latter was hampered by poor infrastructure.

Grenada's percentage annual rate of growth in constant prices for the period 1980-1982 (the year after the PRG take-over and the year before the American intervention) was 5.5 despite continually declining prices for its merchandise exports (World Bank, 1985). This was a better performance than most countries recorded. Moreover, in spite of Bishop's rhetoric, Grenada's direction of trade was fairly diversified. Twenty-two markets accounted for 98% of its exports in 1983. In terms of product and distribution, all Grenada's fresh fruits exports (22.9%), wheat bran (0.9%) and sheep and goats (0.7%) went to Trinidad. Its second leading product export, cocoa beans and waste (22.6%), went to 7 countries which were the United Kingdom, Belgium, Holland, United States, West Germany, the U.S.S.R., and Canada. Its third leading export, nutmegs (18%), went to 26 countries, including West Germany, the U.S.S.R., Holland, and the United Kingdom. Its fourth leading export, bananas (17.9%), went to the United Kingdom, Trinidad, and Barbados while its fifth and sixth, clothing (9.8%) and mace, cloves, spices and condiments (5.3%), went to 12 countries each including Holland, France, the United Kingdom, and West Germany. Thus, for a small country, it was trading with a fairly large number of countries.

A top official in the Ministry of Trade before the election of the present government indicated that there were another twenty markets identified as potential ones for development. One wondered about the choice of markets and strategy, for example, market penetration versus market expansion. This is particularly important since in 1982 and 1983, prior to the U.S.A. intervention, the Communist countries (U.S.S.R., East Germany, Bulgaria, Algeria, Czechoslovakia and Cuba) took 21% nutmegs and 3% mace and 27% nutmeg and 4% mace, respectively, of Grenada's total exports of those products.

Grenada has had certain long-standing commercial relationships, especially with respect to its major crops of cocoa, nutmeg and bananas. Previously, the price of bananas was negotiated with Geese Industries of the U.K., but now, the author was informed it is dependent upon actual sales in the U.K. market. This price is therefore dependent also upon the volume exported and the amount of fruits on the market. To manage the market, fruits have sometimes been left behind or Geese has supplied larger or more ships. Although the contract with Geese Industries of the U.K. does not preclude the Grenada Banana Association from investigating other markets, the banana trade is such a highly specialized and unique trade that the Grenada Banana Association has never tried to extricate itself from the only contract in the banana business that is quite different from the normal banana contract. Either party must give the other three years' notice of intention to alter the relationship.

The export prices per pound of cocoa, nutmeg, mace and minor spices all declined continuously from 1979 to 1983. Cocoa dropped in price per pound from $1.88 U.S. in 1979 to $0.83 U.S. in 1983; nutmegs' prices dropped from $0.88 U.S. in 1979 to $0.59 U.S. in 1983; and mace and minor spices' downward slide in prices were similar to cocoa and nutmegs. All these declines in prices, while the prices for its imports rose, point to the problems of the ever widening gap in the terms of trade between developed and developing economies which are generally resource based and export dependent.

In the case of promotion, there was little or no attempt and concerted effort made to differentiate and promote Grenada's export products, notwithstanding that the Grenada Cocoa Association was created by the Cocoa Industry Ordinance Number 30 of 1964 to safeguard and promote the interest of the cocoa industry and, in particular, to market cocoa and to regulate and control its exports. As in the case of cocoa, all nutmegs are exported through the Grenada Cooperative Nutmeg Association. All bananas must go through the Grenada Banana Association which was set up strictly for the marketing of bananas, while all minor spices go through the Minor Spices Cooperative. Thus, with the exception of the hucksters who look after the selling of their fresh fruits to Trinidad, and some independent merchants engaged in exporting or re-exporting items other than cocoa, nutmegs, bananas and minor spices, the large marketing institutions are statutory bodies established through special acts of the legislature for marketing the specific products. In a modern world of large export trading companies, one wonders about the synergy to be gained from merging these four organizations.

While the design and management of an effective marketing system would be very important for the successful export of the basic commodities of a large number of developing countries such as Grenada, a major bone of contention is the role of government. Under Maurice Bishop's government, Grenada had moved toward a mixed private/public ownership and increased government controlled and directed economy. Most of Bishop's economic initiatives had been aimed at creating a strong state sector of the economy which could generate profits for social programs. Bishop's People's Revolutionary Government allowed the continuation of private banks but set up alternative public banking services to be more responsive to the people's needs. The Grenada Development Bank's (GDB) credibility was evidenced by its ability to raise outside funds for lending which the previous Gairy government was unable to do. In addition to the GDB which channelled all of its loans toward development purposes, the publicly owned National Commercial Bank loaned half its funds for development projects and half for the use of the established commercial sector. The two banks were part of a strategy for stimulating the private sector for it to continue its important role in the economy. The government attempted to provide public subsidies and financing for private enterprise and at the same time sought private financing for public development projects. Its policy was to look locally for development capital before approaching international donors. This approach was based on "moral suasion" (Epica Task Force, 1982, p. 75).

The government attempted to stimulate small and medium-sized businesses while placing some controls on larger ones. Grenada's heavy dependence on imported goods made the commercial sector very powerful and sometimes they exploited the population through higher prices and contrived shortages. Thus the PRG removed rice, sugar and cement from private control and gave sole importing rights for these commodities to the Marketing and National Importing Board, a quasi-public body. It held the prices of those vital necessities down to a level the public could afford and the increase in retail prices over the years were lower than in many developing countries (see Figure 9.1).

The cooperative sector was still in its early growth stage in 1983 when Bishop was overthrown. The cooperatives were not state entities. They were private ventures of the individuals in each co-op, with startup financing from the state. The state loaned money to the co-ops because it was part of a development strategy in which the private sector had shown no interest but which the PRG had considered fundamental to its goals. Thus, Grenada, under Maurice Bishop, had set priorities and made choices according to its social, economic and political goals. In this context the PRG recognized the importance of integrated planning which could boost productivity, for example, by creating links among agriculture, agro-industry, fishing, livestock raising, forestry and tourism. Illustrative of that was to be the use of the waste from the agro-industry and the fish-processing plants for producing animal feed and thereby reducing the cost of raising livestock and by extension, the cost of meat and milk in Grenada. The intervention of the United States in October 1983 brought with it pressures to move to private ownership and market control.

FIGURE 9.1. Grenada: Retail Price Index (1979 = 100)

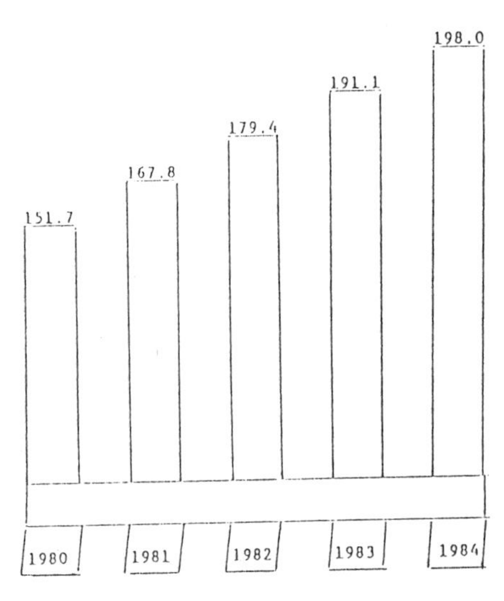

Politics and Marketing After the Intervention

A little more than one year after the United States' intervention in Grenada, parliamentary elections were held. A coalition of three parties formed the New National Party under the leadership of Herbert Blaize, a former prime minister, and won all but one seat in the 1984 elections. Thus, Prime Minister Blaize maintains: "The Government of Grenada has received a mandate from the people to chart *a course for economic recovery and growth which will fulfill the objectives and goals of economic and social development* in the years to come." Further, he states:

> An assessment of the current situation indicated *a need for change in policy in two major areas*. First, it was felt that there should be *greater private sector participation* in economic development; second, it was determined that a *major overhaul of the fiscal system* was necessary to create the appropriate conditions for economic growth. (Investing In Grenada, 1986, p. 1)

Grenada's mission and objectives under Blaize are clearly different from those under Bishop. There is not the emphasis on economic self-reliance and social development programs or on the alleviation of poverty, inequality in income distribution and economic and social injustices found in Bishop's. Under Blaize, Grenada has stopped taxing wealth and started taxing consumption on the advice of U.S. consultants. Prime Minister Blaize has scrapped taxes on personal income, exports, and several other items. He has replaced them with a valued-added-tax, similar to a sales tax, of 20% and a tax on unused land. According to U.S.A. President Ronald Reagan's administration, giving a few people the opportunity to get rich is the best way to raise the standard of living of a country's people (Knox, 1986, p. B12). Grenada's poor who make up the bulk of its 95,817 (1985) population did not have to pay income tax previously but now pay the Value-Added-Tax (VAT) every time they make a purchase except on exempt items such as some unprocessed foods. However, people have suspected merchants of charging VAT on exempt items. After protest from local manufacturers, Prime Minister Blaize, who is also the Finance Minister, has reduced the VAT charged on some of their products. Critics of VAT have maintained that it hurts the poor the hardest and it seems almost designed to keep them in their place. One U.S. official said: "There is no secret about it, a few people are going to get rich, but that doesn't bother us at all. That's the way more jobs will be created and the economy in general will improve" (Knox, 1986, p. B12).

The U.S. Government Agency for International Development (AID) has been instrumental in the flow of more than $70 million (U.S.) into Grenada in less than three years. Many Grenadians have acknowledged the most visible effects of aid such as better roads, water, electricity and other services; however, questions have been muted about the tied aid relationship and conditions attached to the granting of it. AID officials have insisted that the final decision on the package was up to the Government. They maintained that the tax reforms and others, the encouragement to sell state owned companies and farms, raise the ceiling on interest rates from 12.5 to 16% and provide incentives to investors followed policy dialogue associated with U.S. economic support funds (Knox, 1986, p. B12). There were great expectations!

Now that new investment, job creation and commodity production have lagged behind expectations, the economic problems are expected to increase the divisions in Prime Minister Blaize's New National Party. There have been some large scale dismissals of civil servants recently and the resignation of some ministers. The questions that are being raised now are about the Government of Prime Minister Blaize rather than U.S. assistance on which it relies (Knox, 1986, p. B12).

The Economic Profile

The economic profile and marketing in Grenada are changing. The new government early in 1985 established The Industrial Development Corporation (IDC) and in its first year the IDC has approved over 100 projects, including a very significant number of proposals sponsored by Grenadians both at home and abroad. The Government has privatized 18 state owned enterprises and created the National Economic Council to set social and economic development policy and to define strategies for reaching specific objectives.

A comparison of Value-Added Output in 1983 and 1985 is shown in Table 9.1. It reflects some of the problems of the New National Party Government. It also illustrates some of the changes and potential marketing impact, for example, the decline in agriculture and growth in the industry sectors. Other important factors and con-

TABLE 9.1. Comparison of 1983 and 1985 Value-Added Output

Category	1983 Value-Added		1985 Value-Added	
	U.S. $ Million	%	U.S. $ Million	%
Agriculture	18.5	22.8	13.7	19.6
Industry	13.5	14.3	17.6	25.1
Other	52.1	62.9	38.7	55.3
Total	84.2	100.0	70.0	100.0

Source: World Bank and Investing in Grenada.

sideration can be gleaned from Table 9.2, which shows Grenada's economic profile in 1985. The continuing importance of agriculture should be noted as should the potential for development of tourism.

Bennett (1967, p. 233), in his study of Latin America two decades ago, had found that it was important that the overriding economic policy be understood in the context of political institutions and the legislative structure in the environment. The requirements

TABLE 9.2. Grenada: Economic Profile

The most recent estimates of Gross Domestic Product (GPD) are:

	1985 Millions of EC $ (1980 prices)	As % of GDP
Total GDP	188.6	100.0
Agriculture	37.0	19.6
Quarrying	2.0	1.0
Manufacturing	12.0	6.3
Electricity and Water	3.7	2.0
Construction	13.1	6.9
Transport and Communications	14.2	7.5
Hotels and Restaurants	12.2	6.5
Wholesale and Retail Trade	35.4	18.8
Finance and Housing	12.2	6.5
Government	38.2	20.3
Other Services	8.6	4.6

Agriculture is the mainstay of the economy. In 1985, it contributed 19.6% to GDP, about four-fifths of which is attributed to crops and one-fifth to livestock, fishing and forestry. Agricultural exports form 93.3% of total exports as follows:

	% of Total Exports, 1985
Bananas	14.3
Cocoa	23.1
Nutmeg	19.5
Mace	6.6
Fruit and Vegetables	29.8
Other non-agricultural	6.6
	100

The manufacturing sector is small in terms of its contribution to GDP. A recent survey of the manufacturing sector classified companies into the following six groups:

Food production	10
Beverage and tobacco	6
Apparel	7
Wood products	5
Building materials	2
Misc. manufacturers	4
	34

of development had imposed certain imperatives on the direction which policy had taken. Public pronouncements and campaign speeches or even better evidence of ideological orientation were not the best guides to actual policy (Bennett, 1967, p. 233). This is also relevant to Grenada and would appear to have some support in the comparison of the programs of Maurice Bishop and Herbert Blaize.

CONCLUSIONS

The governments of the developing countries span the entire spectrum from monarchies to republics. Nevertheless, most people do not realize that the ideological labels may be misleading. They do not necessarily reflect the degree of democracy, individual freedom, or political rights of individuals and business enterprises. Therefore, the author agrees with Endel-Jakob Kolde (1985):

> No theoretical model adequately describes the business environment of developing countries. They are mixtures of many contradictory and co-existing forces without any discernible pattern. In some respects they are capitalistic; in many other respects they are not. Often they exhibit feudal characteristics, but this model, too, is a very poor facsimile of the realities. (p. 150)

It is clear that existing theories have some serious drawbacks in terms of orientation or focus and applicability to different types of societies encompassing different types of cultures, economics and politics. Thus, where elements in the environment do vary, whether among nations or within nations at different times, it is necessary in assessing and evaluating performance to relate them back to mission and objectives.

While there are some similarities, there have also been major differences in the mission and objectives of the leaders, Bishop and Blaize. This may be worth exploring in greater depth in another study that considers the power of politics and marketing in the Caribbean with their economic and political history and their implications.

The external forces of politics and culture cannot be separated from marketing in the Caribbean. This was recognized almost a quarter century ago when Waldo (1964) perceived an increasing "politicization" of areas of business and the market, resulting from the twin forces of nationalism and socialism. Although Likert maintains that culture is not itself a basic Principle of Management, he asserts, nevertheless, that it influences the method of application of basic principles of management. It appears this was not taken into account by advisors to the Grenada government so that there are a number of problems with respect to implementation of what might appear to be good programs.

Whoever the leader in Grenada might be, any adjustments to its marketing thrust that would rectify the imbalances in trade would require intelligence, creativity, innovation, and skilled political and economic initiatives. The initiatives of Blaize's government are very different from the past Bishop government. Whereas the past government's approach was one aimed at political independence and non-alignment and a controlled mixture of ownership and marketing institutions, the present government's approach is one based on U.S. alignment, deregulation, privatization, and the provisions of incentives for investments.

Perhaps Herbert Blaize, the Prime Minister, with or without pressures from the Reagan Administration, supports Levi's and Dexter's (1983) contention that government involvement is an obstacle to the economy and that this is contrary to the presumed "public interest." Nonetheless, it appears quite clear that the United States' offer of assistance in the areas of tourism, agriculture and light industry which is non-competing with its interest will not be a panacea especially with the insistence of its total capitalist approach. Size, history, stage of development, political, economic, cultural and social factors must be taken into consideration in devising the most effective program for survival, growth and development of Grenada and its inhabitants.

REFERENCES

Bartels, Robert, (1968). "Are Domestic and International Marketing Dissimilar?" *Journal of Marketing* XXX11, July, pp. 55-61.

Bennett, Peter D. (1967). "Marketing and Public Policy in Latin America," in Moyer, Reed (ed.), *Changing Marketing Systems: Consumer, Corporate and Government Interfaces*, Washington, D.C.: American Marketing Association, December, pp. 233-238.

Caribbean/Central American Action, (1986). *Investing In Grenada*. (1986). p. 1.

Drucker, Peter F. (1971). "Marketing and Economic Development." in Taylor, Jack L. and Robb, James F. *Fundamentals of Marketing: Additional Dimensions Selections from The Literature*, New York: McGraw-Hill Inc., pp. 2-10.

Epica Task Force (1982). New York, p.75.

Hettich, Walter, (1983). "The Political Economy of Benefit Cost Analysis," *Canadian Public Policy* IX, pp. 487-498.

Knox, Paul, (1986). "Grenada Being Transformed by AID from United States." *The Globe and Mail*, August 25, p. B12.

Kolde, Endel-Jakob, (1985). *Environment of International Business*. Boston: Kent Publishing, p. 150.

Levi, M. & Dexter, A. (1983). "Regulated Prices and Their Consequence." *Canadian Public Policy* IX, pp. 24-31.

Likert, Rensis (1963). "Trends Toward A World-Wide Theory of Management." *International Management Congress Proceedings* CIOS XIII, New York, p.iii.

McCarthy, E. Jerome & Shapiro, Stanley J. (1975). *Basic Marketing.* (First Canadian Edition), Georgetown, Ontario: Irwin-Dorsey Limited, pp. 4-5, 88.

Peterson, R. & Peterson M. (1981). "The Impact of Economic Regulation and Paperwork," *Regulation Reference Working Paper Series*, Ottawa: Economic Council of Canada.

Rostow, Walt W., "The Concept of A National Market and Its Economics Growth Implications." In Taylor Jr., Jack L. & Robb, James F. *Fundamentals of Marketing: Additional Dimensions Selections from The Literature,* New York: McGraw-Hill Inc., pp. 11-18.

Terpstra, Vern, (1983). *International Marketing*. Chicago: Dryden Press, p. 25.

Waldo, Dwight, (1969). "The Respective Concerns of Business Administration and Comparative Public Administration." In Boddewyn, J. *Comparative Management and Marketing Text and Readings,* Glenview, Illinois: Scott, Foresman, pp. 19-27.

Walvoord, R. Wayne, (1983). "Foreign Market Entry Strategy," *Advanced Management Journal*, Spring, pp. 14-21.

Wells, Louis T. Jr. (1977). "Negotiating With Third World Governments." *Harvard Business Review L*, pp. 72-80.

World Bank Report on Grenada (1985). (Unpublished).

SECTION IV: INTERNATIONAL MARKETING STRATEGIES

Chapter 10

An Application of the Political Economy Framework to Fish Distribution Channels

Erdener Kaynak
Gillian Rice

INTRODUCTION

Many Nova Scotian communities are heavily dependent on the fisheries industry. Fishing, representing 6.9% of Nova Scotian net value of production, leads all other Nova Scotian resource industries such as forestry and agriculture. This paper describes the distribution channel system of the Nova Scotian fishing industry. The information provided here was gathered through focus group interviews with key representatives of fish processors, producers and retailers, and the provincial and federal departments of fisheries. In particular, representatives of the Nova Scotia Department of Fisheries, Fisheries and Ocean Canada, Fisherman's Market, National Sea Products, Brookfield Distributors, Associated Freezer, O'Brien's, McKenzie Limited, IGA, Sobey's, Dominion Stores Limited, and Jonah's Seafoods assisted in gathering the data. Additionally, telephone interviews were conducted with fish processors and distributors from Sydney, Cape Breton area of Nova Scotia. The interviewees were responsible for making major policy decisions regarding fish distribution and marketing.

The description of the channel structure and institutions follows the political economy approach of Stern and Reve (1980) and Arndt (1981, 1983). This approach views a social system, such as a distribution channel, as comprising interacting sets of major economic and sociopolitical forces which affect collective behavior and performance. Table 10.1 contains the major variables that should be evaluated in an application of the political economy framework.

Polity refers to the power and control system of an organization or society; economy refers to the productive exchange system of an organization or society; of the transformation of "inputs" to "outputs" (Arndt, 1981). As Arndt (1983) explains, the political economy view means positioning marketing as exchange behavior between social units where each social unit is a political coalition of internal and external interest groups having partly common and partly conflicting goals. A key feature of the political economy approach is its insistence on simultaneous analysis of the polity and

TABLE 10.1. The Political Economy Framework Applied to Distribution Channels

A. Internal Political Economy	B. External Political Economy
DISTRIBUTION CHANNEL	CHANNEL ENVIRONMENT
1. INTERNAL ECONOMY a) internal economic structure (transactional form that links the channel members) b) internal economic processes (decision mechanisms used to determine terms of trade among channel members).	3. EXTERNAL ECONOMY (prevailing and prospective economic environment; the nature of the channel's vertical i.e. input and output, and horizontal markets).
2. INTERNAL POLITY a) internal socio-political structure (power/dependence relationships between channel members) b) internal socio-political processes (the dominant sentiments, e.g. cooperation and/or conflict within the channel).	4. EXTERNAL POLITY (external socio-political system in which the channel operates; distribution and use of power resources among external actors, e.g. competitors, regulatory agencies and trade associations).

Source: Adapted from Stern, Louis W. and Torger Reve (1980), "Distribution Channels as Political Economies: A Framework for Comparative Analysis," Journal of Marketing, 44(Summer), 56-64.

the economy with the focus on the interactions between them (Arndt, 1981).

Therefore the analysis of Nova Scotian fish distribution channels will begin with discussion of the distribution system (the internal economy and the internal polity; see Table 10.1), followed by a description of the channel environment (the external economy and the external polity) and includes some examples of interactions between the polity and the economy. The purpose of this paper is to provide not only an application of the political economy framework but also a detailed description of a channel system. This can be of interest to both academics and practitioners who wish to learn more about different distribution channels. It can also provide the basis for a comparative study of channels.

INTERNAL POLITICAL ECONOMY: FISH DISTRIBUTION CHANNELS

This section contains an analysis of first, the internal economic structure and processes and second, the sociopolitical structure and processes (see Table 10.1) of Nova Scotian fish distribution channels. The products of the industry vary enormously. The major primary products consist of groundfish species such as cod and halibut, and shellfish such as lobster and scallops. In 1987 approximately 22% of the landings were cod. Scallop landings had a market value of approximately $200 million and lobsters contributed $110 million. The products are processed in numerous ways resulting in a wide variety of derivative seafood products. Through processing, the products available to the end user may be fresh or fresh frozen, frozen (cooked and uncooked), canned, smoked or cured. The product, especially before processing, is generally highly perishable and must therefore be distributed or processed exceptionally quickly. Thus, expedient processing or short, rapid distribution channels are required in order to ensure product quality. In most cases, the end form of the product determines the type and length of the channel that will be utilized (Sims, Brown and Woodside, 1978).

Internal Economy

A summary of how the political economy framework as a whole can be applied to fish distribution channels is given in Table 10.2. The discussion in this section focuses on the analysis of the internal economy. In order to determine the internal economic structure of the Nova Scotian fish distribution system, each of the major channel members is categorized according to its dominant role within the system (see Figure 10.1). Several channel members perform more than one, or overlapping functions; in these cases, vertical integration exists.

Processors

There are over 200 fish processors in Nova Scotia. The two major types are small family operations and large vertically integrated fish processors such as National Sea Products (see Figure 10.1). The small processors usually deal directly with the local retailers or indirectly through small independent buyers who sell to larger wholesalers. These larger wholesalers often collect many small catches and transport them in bulk to institutional and larger retail accounts. Facilitating functions that the small processors perform include activities such as cleaning, filleting, pickling, drying and smoking.

The dominant channel member in the processor category is National Sea Products which is engaged in the landing, processing, storage and physical distribution of many types of sea products. National Sea Products contracts with most of its own buyers and provides the majority of their preliminary storage requirements. Product movement (how and to which markets) is determined by the processing category (canning, smoking, freezing, etc.).

Wholesalers

Fish wholesalers in Nova Scotia vary according to size and activity depending on the type of product handled, their customers, and the facilitatory functions performed. The different types of wholesalers are categorized in Figure 10.1. The smaller regional operators tend to deal with small institutions, corner stores and/or restau-

TABLE 10.2. Applying the Political Economy Framework to Fish Distribution Channels

DISTRIBUTION CHANNEL	CHANNEL ENVIRONMENT
1. INTERNAL ECONOMY a) *small family procuring operations to regional or larger wholesalers to retailers. *large vertically integrated operations (producer-processor--wholesaler) to retailers. *vertically integrated wholesale-retail operations. *independent buyers, void of any contractual ties, form a link between small fishermen and wholesalers. b) *informal relationships among firms; verbal agreements and reciprocity common. *formal contractual relationships within vertically integrated operators, e.g. National Sea	3. EXTERNAL ECONOMY *recession (1981); price decline. *variations in fish stocks. *consumer need for improved quality leading to more direct channels. *growth in export markets for fresh fish.
2. INTERNAL POLITY a) *power of retailers over producers; retailers must agree to stock products. *wholesalers have power to regulate price. b) *cooperation among wholesalers with reciprocal relationships regarding supply. *cooperation between producers and retailers because of retailer power. *conflict between independent buyers resulting from price competition related to the final market, whether domestic or export.	4. EXTERNAL POLITY *fish treaties and quotas (international agreements) *government regulations e.g. entrance requirements, registration and licensing. *government funding of newcomers to the industry, resulting in conflict among channel members. *Canadian National's shipping restrictions. *Air Canada's development of channels for fresh fish distribution.

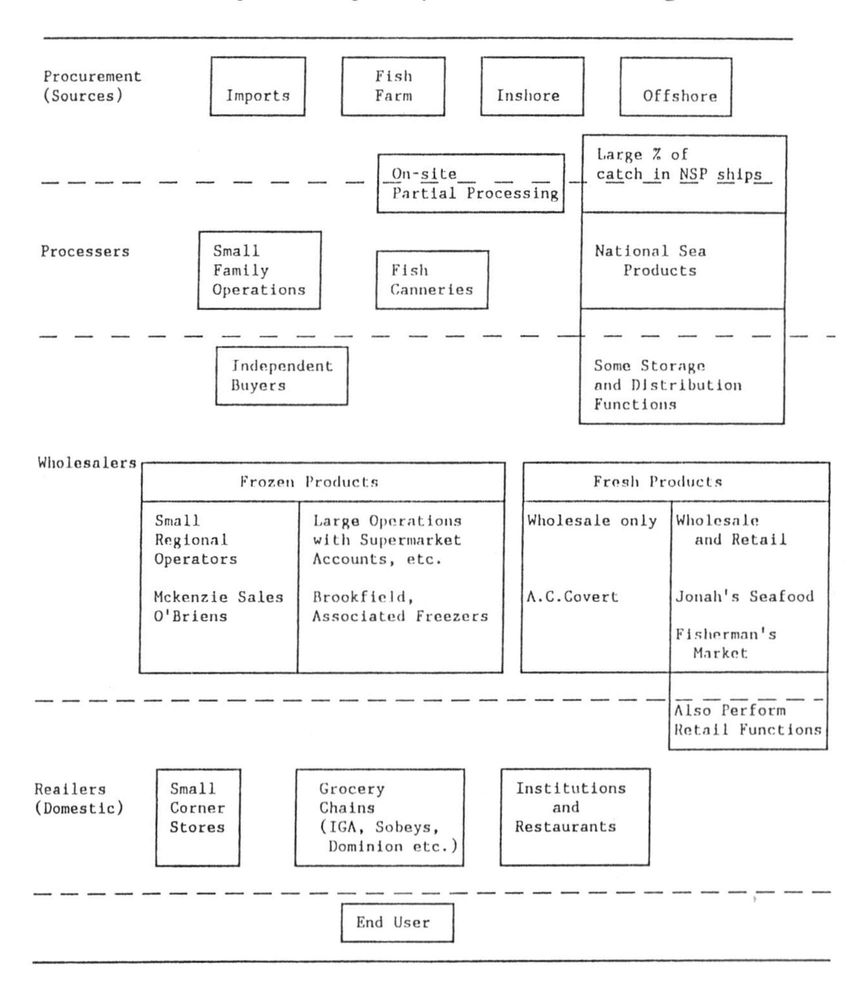

FIGURE 10.1. Overview of the Structure of the Fish Distribution Channels

rants. The Large Frozen Product Wholesalers distribute primarily to the major food chain stores. At times Associated Freezers also rents freezer space to companies such as Swift and National Sea which utilize the facility primarily for the storage of sea products for overseas markets.

The major differences between the Large Frozen Product Wholesalers and the Small Regional Operators (Frozen Products) lie in the scale of operations and the size of the sales force. The smaller operators depend more heavily on an active sales force in order to serve and secure the smaller accounts, restaurants and small institutions. Their basic tasks involve selling new products, reordering and handling customer complaints. McKenzie Sales and O'Briens (see Figure 10.1) both operate in the Halifax-Dartmouth area and offer equivalent services to similar markets. Neither firm handles fresh products of any sort; each has storage facilities and a fleet of trucks.

Wholesalers of fresh seafood products distribute either regionally or locally. These wholesalers can best be categorized by their functions. Some members are exclusively wholesalers while others perform both retail and wholesale operations (some also offer brokerage services). Dominant channel member within the wholesaler-only category, A. C. Covert, deals both with National Sea Products and with smaller independent operations. Covert procures products throughout Nova Scotia, distributing them through its warehouse in Halifax and via its own truck fleet to major supermarket chains such as Dominion and Sobeys.

Of the organizations that provide integrated wholesale and retail functions there are two dominant firms: The Fisherman's Market and Jonah's Seafood (see Figure 10.1). The Fisherman's Market receives products from small producers and through fresh fish wholesalers such as A. C. Covert. As a wholesaler, the firm services some institutional (for example, Camp Hill Hospital) and restaurant accounts. The trend for many large institutions has been a move away from fresh seafood products toward more convenient frozen seafood products. This market is now served mainly by National Sea Products and frozen fish wholesalers rather than by The Fisherman's Market as it had been in the past. As a retailer the firm directly services consumers at the outlet on the waterfront.

Jonah's has two local outlets in Halifax and Dartmouth. The Dartmouth store serves solely as a retail outlet while the Halifax store undertakes some processing, wholesaling and delivery in addition to serving as a retail outlet. Jonah's obtains its fresh sea products from many local independent fishermen. The firm's processing functions include freezing, filleting and some packaging. As a wholesaler the firm services the smaller restaurants and institutions as well as outlets such as fish and chip restaurants.

Retailers

The retail level includes small corner stores and grocery chains, institutions and restaurants (see Figure 10.1). Large chain outlets like Sobey's, Dominion, Save Easy and IGA all handle both fresh and frozen seafood products. The smaller retail outlets deal primarily with small quantities of frozen products and usually only market a partial line (e.g., fish sticks or frozen fillets). These products are usually supplied by small regional wholesalers like McKenzie in Dartmouth or O'Briens in Halifax.

Transactional Forms and Channel Linkages

To summarize the above description in terms of transactional forms, the internal economic structure of the fish distribution channels comprises two sub-systems. One system includes the small family producing operations, independent buyers, regional wholesalers and smaller retail outlets. The other includes large vertically integrated operations covering production, processing, wholesaling and retailing, dominated by National Sea Products. Vertical integration also exists in the wholesale-retail operations of The Fisherman's Market and Jonah's but a much smaller scale than that of National Sea.

The system of independent channel members and the vertical marketing system of National Sea Products are closely linked. The major linkages can be described in terms of those for frozen products and those for fresh products. With respect to the former, National Sea is the major producer of frozen seafood products, with supplies procured from its own offshore fishing fleet and from independent inshore fishermen via the independent buyers (see Figure 10.1). Following processing and storage, all distribution of the product is achieved indirectly to retailers through Brookfield and Associated Freezers. These wholesalers perform not only the facilitating functions of storage and transportation but also the function of sorting. The wholesalers generally hold huge inventories and when an order is received, delivery to a retailer can take place within twenty-four hours.

Fresh products are distributed in a similar manner but with the fish being transported from National Sea Products or small independent operations to A. C. Covert. Retail outlets, again, base their ordering on turnover and demand and each independent outlet generally orders from the wholesaler. Because of the fact that the quality of the product may deteriorate rapidly, orders are usually filled on the same day and retailers generally place orders daily.

Dominion Stores attempted a conversion from a decentralized distribution policy to an experimental centralized policy. The centralized policy meant that the individual stores procured their products through a central regional office. Previously, each individual store had ordered their own products when demand dictated. The new system, however, soon proved ineffective for two reasons. First, the orders were being processed only twice a week instead of daily as required and therefore the system did not meet the demand requirements of the retailer. Secondly, the resulting product quality was inferior due to prolonged holding times and the perishable nature of the product. Dominion has since reverted to the decentralized policy with each store once more in charge of its own needs and requirements.

Jonah's and Fisherman's Market, the wholesaler-retailer organizations, generally secure their products from independent community buyers or from A. C. Covert. The end user must generally pick up the product unless the account is a large one; in this case delivery would be made.

Processes

The internal economic processes of the fish distribution channel include both formal and informal arrangements. While National Sea Products has contracts with some other channel members it is also clear that while other arrangements may be more informal they are conducted on a very regular basis. Because of the nature of the product (extremely perishable), Dominion's experiment with centralized ordering shows that decentralized transactional arrangements predominate. Nevertheless, it seems that the division of functions among channel members (processing, storing, distributing, etc.) is routine or habitual and is based on long-term relationships.

More information needs to be gathered about the internal economy to determine to what extent the fish distribution channel represents an administered market with transactions occurring inside the boundaries of a group of companies committed to long-term cooperation (Arndt, 1979). In his discussion of "domesticated markets" Arndt (1979) explains that the relationship between the companies involved may acquire its stability from implicit agreements, understanding and subcontracting arrangements. As will be explained in the following section on internal polity, verbal agreements and reciprocal arrangements are common between wholesalers.

Internal Polity

The internal polity of a distribution channel comprises the internal sociopolitical structure and processes (see Table 10.1). Issues here are the power and dependence relationships between channel members and cooperation and conflict within the channel. This study of distribution channels did not include collection of data for the formal measurement of power, dependence, cooperation or conflict. Nonetheless, from the information gathered during the informal interviews it is possible to speculate on these variables in the context of the political economy framework.

Power and Dependence Relationships

Power is "the capacity of a particular channel member to control or influence the behavior of another channel member" (Rosenbloom, 1973). Among the "behaviors" more powerful channel members may seek to control are elements of other members' marketing strategies. There will always be *some* power existing within channels due to mutual dependencies which exist among channel members, even though that power may be very low (El-Ansary and Stern, 1972). Etgar (1976) suggests that because of the numerous marketing flows which link channel members together, the more common case is a mixed power situation where different firms exercise control over different flows, functions or marketing activities.

Two examples of this in the case of the Nova Scotian fish distribution channels concern pricing and retailer acceptance of products. The wholesaler usually regulates the market price by noting what the end consumer will pay, working it down to the fisherman and up to the retailer. Thus, the fisherman and the retailer are generally price takers and the wholesaler is the price maker. The prices to fishermen vary from district to district depending on proximity to the market, the season, and the size of the catch. The retailer is faced with the problem of trying to equate supply and demand. The wholesaler usually has the ability to process, for example freeze, the product if there is a glut; the retailer often does not have this capacity.

The relationship between price and channel power is also illustrated by the independent buyers who make pick-ups from many small producers and sell to the centralized wholesalers. Independent buyers are void of any contractual ties and often provide fishermen with financial assistance in order to secure their supply. The fishermen are often dependent on these buyers as a link to the market place, as it is not economical for them to sell directly to the end user. The wholesaler usually bears the transportation costs incurred by these buyers. The buyers, especially in the case of lobster, will attempt to hold the product if they feel the price will increase during that time.

Retailers have the power to accept products. For example, if National Sea wishes to list a new product or brand, the firm must contact the large retailers in order to secure their cooperation. If the retailer does not wish to carry the product National Sea does not have the power to force it on the retailer. If the retailer agrees to stock the product or brand, he then goes directly to the wholesaler who procures the product from National Sea and distributes it to the retailer. Note that while the product flows remain the same, the communications channels (for example, between producer and retailer) differ from those utilized for the more usual transactions.

These two examples of power seem consistent with Patti and Fisk's (1982) conceptualization of how channel dominance for the wholesaler is based on price competition, while manufacturer and retailer dominance is achieved with a combination of advertising and branding.

Cooperation and Conflict

The internal sociopolitical structure of a channel is closely linked to the internal sociopolitical processes. The processes refer to the dominant sentiments such as cooperation or conflict within the channel. McCammon (1970) discussed how conventional marketing channels, comprised of isolated and autonomous decision-making units, are unable to program distribution activities. The relationship between power and cooperation is such that if power is low, so is dependence, therefore two or more relatively independent entities may not be motivated to cooperate (Stern and Reve, 1980). In the Nova Scotian fish distribution channels examples of cooperation and conflict are found. The dominant sentiment between the retailer and the producer is cooperation. This seems to be related to the power of the retailer with respect to the stocking of goods.

An interesting case of reciprocity exists between fish wholesalers in the Metro Halifax area (Halifax, Dartmouth and Bedford). It is not unusual for a great deal of cross-shipments of products to occur. For example, A. C. Covert may supply the Fisherman's Market one day with the reverse occurring on another occasion. Delivery can usually be conducted within hours from one wholesaler to a "competitive" wholesaler, should the latter become stocked out. The fluctuation in demand and the perishability of the product create the need for the utilization of this procedure. In terms of economic processes, this type of agreement is not contractual but is simply made on an ad hoc basis initiated and sealed solely through verbal, usually telephone, negotiations. In one interview conducted for this study, one of the wholesalers stressed that a participant in the reciprocal agreement need only to violate the verbal agreement once and from then on the party at fault would seldom be trusted or aided in future negotiations. The case of reciprocity illustrates how an economic process, that is, reciprocity among the fish wholesalers is associated with a relationship of dependency (an internal political state) specific to an industry where the products are extremely perishable. Such interactions between economic and political forces are important. Stern and Reve (1980) emphasize that the essence of the political economy framework for the analysis of marketing systems is that economic and sociopolitical forces are not analyzed in isolation.

Some years ago there was a case of conflict between the smaller, independent buyers and Nickerson, an organization that was outbidding the small buyers. The latter felt that it was in order to eliminate their competition. Nickerson was, however, dealing with the European market which paid higher prices and therefore could afford to pay fishermen higher prices than could the buyers relying on the domestic market. This did have the effect of eliminating some small buyers, nevertheless.

EXTERNAL POLITICAL ECONOMY: THE CHANNEL ENVIRONMENT

It has been shown empirically that organizations depend on their environments (for example, see Pfeffer and Salancik, 1978). In this section, the discussion focuses on some important variables in the channel environment of the Nova Scotian fish distribution system. Using the political economy framework these variables are divided into two groups: those relating to the external economy and those relating to the external policy. Illustrations of these variables with respect to fish distribution are given in Table 10.2.

External Economy

The external economy includes the prevailing and prospective economic environment and the nature of the channel's vertical (input and output) and horizontal markets (see Table 10.1). In the early eighties the Atlantic fisheries were the most troubled industry of Canada's most troubled region (Copes, 1983). The industry was badly affected by the recession of 1981 when prices declined. Currently, however, prospects for the industry are much improved. Ellison (1985) reports the growth of export markets for fish, especially for high quality and specialty seafood products.

In terms of inputs to the fish distribution channels, interviewees stressed the vulnerability of the industry to weather conditions. This uncontrollable variable makes scheduling difficult. Yearly variations in the size of fish stocks themselves also can affect the industry. Depending on whether a glut or scarcity is experienced, channels members must cope with the problems of physical movement or supply.

One area which appears to require attention is the need for improved quality. Industry representatives hope that better quality will ensure continued market penetration and growth along with higher prices for the processors and fishermen. As the industry becomes more competitive, surviving members will be those with the best quality. Stress on quality could have an impact on the present distribution channels. Since the product is highly perishable, ways of shortening the channel between processor and eventual consumer may need to be developed. This perhaps can be done through vertical integration—that is processor owning and having more power over wholesaling function.

External Polity

In the last decade, because of events in both the external economy and external polity, the Nova Scotian fish distribution channel has faced a turbulent environment. One important event was the delay by the United States to ratify the George's Bank fishing treaty at the end of the seventies. This was an issue taken to the World Court in The Hague before settlement was achieved. At the time, Nova Scotia's share of the George's Bank scallops (one of Nova Scotia's most important species) dropped from 90% to 60%. This decline was blamed directly on the massive build-up of the U.S. scallop fleet which was free to harvest as many scallops as possible. Today, a significant issue facing the fishing industry is whether or not oil-drilling will be permitted on George's Bank. There is interaction between the economic success of the channel members in the fishing industry and the completion of international agreements and setting of quotas. A political decision permitting oil drilling on George's Bank could have some effect on the location of the fish feeding areas and the variations in fish stocks.

Government regulations and interventions have also had an impact on the supply of fish and the activities of channel members. Seasonal restrictions, quotas, entrance requirements, registration and licensing tend, at times, to protect, but at other times to restrict the efficiency of operations. The government policy of funding operations has sometimes been discouraging to certain industry representatives. Frequently the older established operations are unable to obtain government support while newcomers take full advantage of it. This causes a situation where newcomers to the industry compete directly, with the help of government funds, with the established firms whose successful operations may be damaged. Thus government intervention results in conflict among channel members.

A short-term problem of distribution that could have affected channel arrangements but has been solved occurred when Canadian National Railways limited its shipping capacity for fish with respect to days of shipment. Canadian National (CN), for a time, reduced its days of shipment from six to two days per week. In addition, an increase in free structure made shipment expensive even to local areas unless the shipments were made in large quantities (over 500 lbs.). The physical distribution was also changed to CN truck rather than CN rail. A considerable amount of stress was felt by producers and wholesalers who utilized CN's transportation facilities. The prohibitive cost of air transportation leads to relatively little of the product being moved in this way.

Air Canada, however, in a development unrelated to the CN problem, has played an active role in increasing the international distribution of fresh fish by air freight. The interesting fact, from the political economy viewpoint is that, as Ellison (1985) explains, Air Canada has actively developed the channel for fresh fish distribution, rather than adopting the traditional role of being a facilitating agent. It has done this by utilizing two power bases: expert and reward. Air Canada has expertise in air cargo and has been able to increase shipments by participation in international food shows and exhibitions, affording it broad exposure to potential markets. The airline also contacted potential fresh fish shippers to inform them about its service and how they could be rewarded by using these services to ship fish to distant domestic and international markets.

In postulating causal formulations within the political economy framework, Arndt (1983) suggested that a common pattern is that the internal polity is affected by its relations to the external polity (and to some extent, to the relations to the external economy). In turn, the internal polity transmits rules and allocates resources to the internal economy. This can be illustrated in the context of the fish distribution channels as follows. The growth in export markets for fresh fish and specialty seafood (external economy) has contributed to the active development by Air Canada of the air freight distribution system to serve these markets (external polity). Both of these factors have contributed to increased conflict from price competition among intermediate members of the fish distribution channels because of the higher prices possible in the export markets (internal polity). This can result in a smaller number of independent buyers in the channel (internal economy).

CONCLUSION: THE FUTURE OF FISH DISTRIBUTION CHANNELS

To an increasing degree, transactions are occurring in "internal" markets within the framework of long-term relationships, not on an ad hoc basis (Arndt, 1979). However, in a comparative analysis of the domestic retailing systems of Canada and Sweden, Kaynak (1985) found that vertical marketing systems are not strongly prevalent in Canada. He reported that the implementation of forms of vertical marketing systems in Canada should contribute to the increased efficiency of the Canadian retailing system. Indeed, Stern and Reve (1980) argued that the more that the exchange process among channel members is organized, the more efficient they will be. As environments have become more turbulent there is an increasing need for firms to buffer environmental influences by gaining more control of supply and demand (Arndt, 1979).

It can be argued that this is the case in Nova Scotian fish distribution channels. Not only is the environment turbulent, but the distribution system itself is extremely complex. Most of the distribution functions overlap in the categories of processor, wholesaler and retailer. The system, because of the perishability of the product lines, is characterized by a great deal of ad hoc arrangements and cooperation. This has been illustrated by the political economy analysis of the fish distribution system. In the future it seems likely that there will be more vertical integration from the wholesale level. Wholesalers will integrate backwards by employing dependent community buyers and will integrate forwards toward the retail level. Already, the data collected for this study show that some of the independent buyers act as representatives of purchasing agents for large vertically integrated organizations. These organizations could achieve increased channel power and economic efficiency through ownership of the independent buyers.

To conclude, the analysis of the Nova Scotian fish distribution channels using the political economy framework improves understanding of the channel members and their linkages to one another in economic terms and sociopolitical terms. The examination of both the economy and the polity highlights the interaction between them and can help in suggesting how developments might occur in the fish distribution channel in the future.

REFERENCES

Arndt, Johan (1979). "Toward a Concept of Domesticated Markets. "*Journal of Marketing*, Vol. 43 (Fall), pp. 69-75.

______ (1981). "The Political Economy of Marketing Systems: Reviving the Institutional Approach." *Journal of Macromarketing*, Vol. 1 (Fall), pp. 36-47.

______ (1983). "The Political Economy Paradigm: Foundation for Theory Building in Marketing." *Journal of Marketing*, Vol. 47 (Fall), pp. 44-54.

Copes, P. (1983). "Fisheries Management on Canada's Atlantic Coast: Economic Factors and Socio-Political Constraints." *The Canadian Journal of Regional Science*, Vol. VI, pp. 1-32.

El-Ansary, A. & Stern, L.W. (1972). "Power Measurement in the Distribution Channel," *Journal of Marketing Research*, Vol. 9 (February), pp. 47-52.

Ellison, Robert A. (1985). "International Air Cargo: Facilitating Agent for the Distribution of Fresh Fish from Atlantic Canada." Shaw, Susan, Sparks, Leigh & Kaynak, Erdener. *Proceedings, Second World Marketing Congress*, August 238-31, pp. 745-754.

Etgar, M. (1976). "Channel Domination and Countervailing Power in Distribution Channels." *Journal of Marketing Research*, Vol. 13 (August), pp. 254-262.

Kaynak, Erdener (1985). "A Comparative Analysis of the Domestic Retailing Systems of Canada and Sweden." *Journal of Macromarketing*, Vol. 5 (Spring), pp. 56-58.

McCammon, B.C., Jr. (1970). "Perspectives for Distribution Programming." Bucklin, L.P. (ed.), in *Vertical Market Systems*, Glenview, IL: Scott, Foresman, pp. 32-51.

Patti, Charles H. & Fisk, Raymond P. (1982). "National Advertising, Brands and Channel Control: A Historical Perspective with Contemporary Options." *Journal of the Academy of Marketing Science*, Vol. 10 (Winter), pp. 90-108.

Persson, Lars (1982). "Evaluation of Retail Systems," Dietl, J. & Banasiak, A. (eds.), *Socioeconomic Effectiveness of Distribution System*, Lodz, Poland, pp. 53-65.

Pfeffer, Jeffrey & Salancik, Gerald R. (1978). *The External Control of Organizations: A Resource Dependence Perspective*. New York: Harper.

Rosenbloom, Bert (1973). "Conflict and Channel Efficiency: Some Conceptual

Models for the Decision Maker." *Journal of Marketing*, Vol. 37 (July), pp. 25-30.

Sims, J. Taylor, Brown, Herbert E. & Woodside, Arch G. (1978). "An Integrative Model of the Channel Decision Process." In Woodside, Arch G. et al., (eds.), *Foundations of Marketing Channels*, Austin, TX: Lone Star Publishers, Inc., pp. 47-65.

Stern, Louis W. & Reve, Torger (1980). "Distribution Channels as Political Economies: A Framework for Comparative Analysis." *Journal of Marketing*, Vol. 44 (Summer), pp. 56-64.

Williamson, Oliver E. (1975). *Markets and Hierarchies.* New York: The Free Press.

Chapter 11

Socializing Grain Distribution in Kongi* State for the International Market

Bedford A. Fubara

INTRODUCTION

According to Terpstra (1978), international marketing refers to that marketing which crosses national boundaries—that which has "inter-nation" aspects to it. We are aware, however, that most firms in developing countries are engaged in export sales of their goods in the international marketplace and such firms are generally state-owned because private firms in those countries more often lack the ability to engage in international competitive strategies. These state-owned enterprises (SOEs) with the backing of their national governments, employ strategies which are peculiar to them in distributing their goods abroad. It is, therefore, not easy to say that SOEs engage in international marketing since that would seem to assume that they are in international business. Since international business involves investment overseas in factory facilities for assembly or manufacture, it may also demand full multinational operations which SOEs are ill-equipped to undertake.

The distinction between domestic sales and overseas sales would seem reasonably straightforward but it is not easy to draw a dividing line between exporting and international marketing or international business. Established authorities (Tookey, 1969; Kotler, 1972) allow the inclusion of direct sales from home markets to foreign us-

*Kongi State is the disguised name of the country where the study was carried out.

ers, distributors or agents and the use of the various forms of joint venturing to establish overseas marketing subsidiaries, as export. Therefore international market and exporting or overseas sales is used interchangeably to mean the same thing.

Product market strategy is concerned with how a firm is to compete in the international marketplace (Piercy, 1982, p. 10). One of the most seductive tenets of SOEs in international market is that they are capable of operating competitively using the non-market grounds in the key market chosen rather than relying on their competencies and price competitiveness. Until recently price competitiveness in export was seen as paramount (Mackay, 1964; Parkinson, 1966) but according to NEDO 1977 recent argument is that price competitiveness may be superceded in power by non-price factors, especially where SOEs are being used as instruments for generating foreign exchange for the nation.

In the next part of this paper, we shall examine the socio-political aspects of grain marketing in Kongi State as a way of explaining the state of its behavior in the international market by looking at (a) the political, economic and operating environments (b) the legal background (c) the marketing strategies and (d) the social implications of the Grain State enterprise operating in the international market.

THE SOCIAL-POLITICAL ENVIRONMENT

Broadly speaking, marketing is the collection of activities undertaken by firms to relate profitability to its market. These include marketing research, product development, pricing, distribution and promotion, usually called marketing mix. In international marketing, it is the performance of one or more of these activities across national boundaries that is called international marketing. The unique dimensions of international marketing are that they involve responding to both "controllables" and "uncontrollables" in the various countries of operation and one can imagine that these factors could not be the same in all countries. Among other things, the characteristics peculiar to each national market would be different, politically, culturally and otherwise.

The Political Environment of Kongi State

Kongi State witnessed some great changes since the eruption of their revolution 14 years ago. According to official documents, the measures embarked upon at the early stages of the revolution created favorable conditions for embarking on a socialist-oriented economy. Essentially the sectors of the economy are under the control of the state enabling the state to employ the surplus generated from economic activities for the improvement of quality of life. The country is said to be undergoing the process of social transformation having the backing of the working people whose philosophy is modelled along the Marxist/Leninist thinking.

In this process the Government has adopted central planning, normally practiced in socialist economies as the main instrument in the management, guidance and acceleration of the socio-economic transformation of the country. The political commitment to central planning was crystallized in the establishment of the national Commission for Central Planning which was set up 10 years ago. So far, six annual plans have been prepared and executed as part of the effort to intimate the social and economic objectives articulated by the national democratic revolution.

The political philosophy is to strengthen the gains of the revolution, namely to free the working people completely from "reactionary" production relations; to withstand and repulse the onslaught of imperialism; to permit working people to enjoy the fruit of their labor and raise their standard of living as well as enhance capital formation which is the key to qualitative and quantitative increase in output in the transition toward a modern socialist economy.

According to their Ten Year Perspective Plan, the aim of the government is to mobilize and direct the country's human and natural resources towards direct productive activities, particularly capital formation, so as to raise the level of development of productive forces and standard of living. It is believed that this plan would propel the country out of the abyss of economic backwardness by enunciating appropriate development strategies. Current claim from official sources is that central planning philosophy has produced encouraging results since its inception because the annual plans have helped to rehabilitate the economic and social infrastructure destroyed by internal and external reactionary forces. It is also claimed that efforts in this direction have helped to minimize the effects of acute economic crises in stabilizing the national economy and laying down a sound basis for its future planned development.

Before the revolution, Kongi State's socio-economic growth was constrained by the pseudo-capitalistic mode of production. The development of the national economy was more or less left to the interplay of spontaneous market forces. The peasantry which made up the majority of the population had no incentive to raise the productivity, because the landlord-tenant relationship deprived them of up to 75% of their produce in addition to lack of security of tenure. Under socialist system ownership of the means of production, the management and development of the economy are squarely left in the hands of the government. The central planning supreme council therefore saw its major objectives as *inter alia* alleviation of acute urgent economic and social problems; the reconstruction of social and economic infrastructure and laying down of material and technical foundations for building socialism. The major objectives of the government therefore are (a) to accelerate the growth of the economy through the expansion of the country's productive capacities; (b) to ensure structurally balanced development of the national economy by expanding domestic resource based industries; (c) to conserve, explore, develop and exploit rationally the natural resources of the country; (d) to expand and strengthen socialist production relations along socialist principles; (e) gradual reduction of unemployment; (f) to ensure balanced and proportional development of all regions of the country by developing each region according to its own comparative advantages of natural resources.

In achieving the above objectives it is also recognized that the priority area is the agricultural sector because agriculture provides employment to a large sector of the population and plays a major role in generating financial surplus and foreign exchange needed for economic development. In fact agriculture has been described as the "foundation of the country's economy."

The Economic Environment

Agriculture is the mainstay of Kongi State economy and will continue to play this role for many years to come. Currently, agriculture accounts for 45% of the Gross Domestic Product (GDP) and

85% of exports and is the means of livelihood of more than 80% of the population.

With the per capita GDP of US $110 – Kongi State is one of the poorest countries in the world and belongs to the least developed country group. Its population is estimated at 42 million. The population density is 34 per sq. km. The area used for agriculture at the moment represents only 49.8% of the total. Out of 5.9 million hectares of arable land about 5 Mha are used for cereals. Ninety-four percent of the land is used by the private farmers, 3.6% by producer co-operatives and 2.5% by State farms. Crop production is estimated to contribute 60% of the agricultural GDP.

Following 1975 land reforms, farmers have user's rights to cultivate land and farm one to two hectares predominately on a subsistence basis. Peasants cultivate about 96% of the crop areas and produce some 90-94% of cereals, pulses and oil seeds with somewhere between 25 to 30% of this production being marketed.

The government has been conscious in its approach to adjusting past agricultural development policies. The resource constraints have been natural disasters. The problem of transforming a deeply traditional society has proved quite difficult in a country with poorly developed infrastructure and a poor, largely agrarian, economy. With a decade of agricultural production stagnation in the face of population growth of about 2.9% p.a., the dream of a poverty and hunger free country would remain an illusion if management policy to increase agricultural growth is not given the first priority. Government has therefore proposed production targets for the next ten years as shown in Table 11.1.

It is also envisaged that with the projection in production, there will be structural changes in the mode of production as shown in Table 11.2.

The prime objective is clear although the targets proposed are optimistic and out of tune with past performances. For instance the plan shows that crop production volume increase should exceed 5% annually, which can only be achieved through substantial crop yield increases and expansion of the crop area cultivated. To achieve the 5% ambition and the structural changes in production which will unlock agricultural potential, requires a policy embracing strategies to market its grain in the international market to attract foreign exchange to pay for imported machinery.

THE LEGAL BACKGROUND OF THE GRAIN STATE ENTERPRISE

In the context of the proclamation setting up the GSE, it is expected that the enterprise would be the major source of socializing grain purchases from the state co-operatives in order to sell locally to government agencies including those agencies instituted for purposes of trading in the international market. Therefore, the powers

TABLE 11.1 Planned Production Projection in Agriculture (in Thousands of Quintals)

	1983/84 Actual	1983/84 Planned	1993/94 Planned	Annual Growth
Cereals	55,268	59,212.4	105,174.1	5.9%
Pulses	7,117	7,561.2	11,299.6	4.1%
Oilseeds	985	1,158.5	2,555.4	8.2%
Coffee		1,179.3	2,772.3	4.5%
Others		25,551.9	44,164.0	5.6%

Source: Government documentations (1988)

TABLE 11.2. Share of Cultivated Land (in Thousands of Ha)

Entity	Area 83/84	Share %	Area 93/94	Share %
State Farms	221.4	3.4	502.2	6.2
Prod. Coop's	88.7	1.4	3,938.8	48.9
Individual Farmers	6,136.8	94.6	3,500.0	43.3
Settle Schemes	39.6	0.6	121.8	1.5
Total	6,486.5	100.0	8,063.8	100.0

Source: Various Government documents (1988)

and duties of the GSE are spelled out in the proclamation setting up the enterprise and are as follows:

1. to buy grain from suppliers and sell to
   a. Mass organizations and other organs engaged in retail trade.
   b. Public Enterprises engaged in export trade and
   c. government offices
2. to supply grain to government, mass organizations and private factories that use same as raw material;
3. to maintain a national emergency grain reserve;
4. to construct, equip and maintain, for its own use, buildings, silos, storage facilities, grain elevators and other structures and machinery;
5. to sell or otherwise dispose of, in accordance with the directives of the Minister, any grain prone to deterioration or unfit for human consumption;
6. to enter into contracts;
7. to sue and be sued in its own name;
8. to acquire, possess, pledge, sell and exchange property;
9. to perform such other activities as are conducive to the attainment of its objectives.

The grain state enterprise was first established in 1960 through a charter but repealed in 1976. The transition of the GSE to its present state was re-established in 1987 by an act of proclamation.

The concept of the GSE as it is today, is to socialize grain marketing in such a way as to "eliminate" competition and to "wither-away" private enterprises in order that grain purchase from co-operatives' producers may be firmly placed in the hands of the state enterprise. In order to achieve this objective, the following strategies are employed:

- Government has fixed the prices of grains at 40% below the market prices;
- State farms have been established to produce and sell to the GSE only at the fixed price;
- Co-operatives and peasant farmers are sanctioned to sell 100% of their production to the GSE at the fixed price;

– Private producers and settlement farms are by law required to sell 50% of their production to GSE but could sell the 50% balance in the open market at higher prices.

To ensure that the farmers co-operatives and private producers sell their grains to the GSE the national government established a national task force (NTF) to police the delivery of the grains to the GSE collection points. Each co-operative society and private producer association has a target that it must meet in every season until off-season. Information available to us during the study is that some private associations and co-operators have been constrained to buy from open market at higher prices to make up their target quantity. In order to ensure that grains are collected for GSE distribution locally and internationally the enterprise has the following establishments.

14	Administrative Regions
104	Procurement and distribution offices in the districts
5	Regional offices
17	Branch offices
105	Procurement Stations
1768	Grains Acquisition Centers

We shall examine the distribution of grains locally and internationally with a view to assessing the viability of the strategy employed in this marketing exercise.

GSE MARKETING STRATEGIES

GSE is the leading monopoly organization in grain procurement and marketing in Kongi State. It has the backing of the ministry of Domestic and Foreign trade and the Ministry of Agriculture. In fact, to ensure that the co-operative societies and state farms produce sufficient grains, the Ministry of Agriculture has an Input Corporation which supplies fertilizers and other agricultural inputs to the societies. The Input Corporation is funded by the World Bank. The Co-operatives are encouraged to pool their resources together to provide storage and transport facilities for themselves in order to provide depots where the GSE can conveniently collect grains. GSE therefore has the following as its sources of grains supplies—see Table 11.3.

As stated earlier, the national task force is expected to ensure that GSE buys the 50% production from private traders and settlement farms at fixed prices. The balance of the grain (50%) is sold in the open market at 40% higher than the fixed price. This means that grains in Kongi State have two prices. The consequence is that producers have resorted to selling low quality grains to the GSE while reserving the higher quality grains for the open market and thereby increasing the chances of post-harvest losses to GSE. Again, what constitutes 100% production of the private producers and traders is not clear because we found that there might be an element of hoarding in the whole arrangement since it is not possible for the NTF to be in a position to prove what quantity of grain constitutes 100% of the production at any point in time.

Arising from the objectives and powers set out in the proclamation setting up the GSE, the enterprise supplies its grains to the following outlets as shown in Table 11.4.

MARKET SOCIALIZATION PHILOSOPHY AND THE INTERNATIONAL MARKET

The whole socialization philosophy in the context of Kongi State is to ensure that the distribution of grains is left in the hands of government and its agency. The agency is to ensure grain distribu-

TABLE 11.3. Grain Procurement Sources

State farms	35%	= 100% Production of the farms
Service Co-operatives	37%	= 100% Prod. of the Co-operatives
Traders	15%	= 50% Production of the Traders
Settlement farms	13%	= 50% Production of the farms
Total	100%	

Source: Interview Data (1988)

TABLE 11.4. GSE Grain Distribution Strategy

1.	Public Industrial Establishment	24%
2.	Basic Commodities Distribution Corporation	17%*
3.	Public Consumer Organisations	6%
4.	Grain Export Corporation	42%*
5.	Urban Centres and Grain deficit areas	11%
	Total	100%

Source: Interview Data (1988)

*Sales to these corporations are made at 50% of the purchase prices of the GSE to socialize grain distribution.

tion locally at low prices as part of welfarism and grain distribution in the international market at a price at which the state grain corporation can compete favorably in order to earn foreign exchange which can be employed in the purchase of technologies for local state industries.

The philosophy of socialization articulates the following aims:

1. To maintain equitable grain distribution by supplying from grain surplus to grain deficit areas
2. To develop a market free from exploitation by speeding up the socialization process
3. To contribute to the foreign exchange earning capacity of the country by supplying the required foreign market agricultural commodities to the public export agencies.
4. To execute government policy with regard to marketing and pricing of agricultural commodities.

A document of the Kongi State government says:

> The socialization model was prepared with the aim of expanding the share of the socialist sector in the wholesale trade. The most important characteristic of the model is to be able to eliminate and exclude the private wholesale traders.

The expectations of the socialization model is to avoid duplication or overlap of functions, extended marketing chains or any competition in the market. "The model precludes any trial and error measures as well as drafting unnecessary plans and programs of action and the subsequent waste of time and resources" (Government statement).

The Grain Export Corporation (GEC) receives grains from GSE at 50% of the GSE purchase price. The government in this strategy has decided to allow the GEC to have a competitive edge over other suppliers of the same grain in the international market. This means the government is prepared to bear the loss locally by paying the price differential to GSE in local currency as subsidies. The reason for selling grain at 50% of GSE price to the Basic Commodities Distribution Corporation is to employ the Corporation in the distribution of food to the starving population.

Most prescriptive theory in the field of internationalization of marketing rests on the assumption that organizations make "rational" decisions, usually in the sense of seeking acceptable goals and evaluating alternative ways of achieving them before making decisions on courses of action. This assumption is perhaps clearest in economic theory (Piercy, 1981). But organizations have gone into international market based on unsolicited orders (Day, 1976). There is a view that some organizations have gone international on reactive grounds. For instance an organization may send its goods to the international market because of excess capacity at home (Arpan, 1972). In another circumstance Robinson (1967) had reported that companies had gone international with their produce because of external pressures and indeed some organizations internationalize their products to overcome declining home demands. In the circumstance of Kongi State, it has decided to go international with its grains because of the desire to be able to pay for its imports of machinery, despite its grain deficiency at home. That decision has more of a reactive overtone than anything else, yet it is strategic.

Assessing the production capability of GSE in financial terms, there is evidence that the enterprise has been operating at residual deficit reserves since 1983. With its present staff strength over 4,000 persons and per staff sales and purchases below 1,000 quintals in 1988, grain production would not be sufficient locally.

Therefore foreign sales would have been out of the question. But since going international is based on external pressures, the enterprise had to go international to cope with the pressure—see staff productivity in per quintals of grains in Table 11.5.

IMPLICATION OF SOCIALIZING GRAIN DISTRIBUTION

From the data available to us we find that the implications of socializing grain distribution in Kongi State include:

(a) Loss of fund in operation, arising from buying below the market price which attracts poor quality grains, and selling at 50% of purchase price to Basic commodity corporation and to the public export corporation which means operational costs are not covered. This seems to create more social problems.

(b) Undefined operational scope of the enterprise which strains its operational ability and therefore exposes its operational results to ridicule, because the number of staff employed (4152) and the GSE administrative and organizational spread do not justify the poor performance of less than 1,000 quintals per capita.

TABLE 11.5. Staff Productivity (in Quintals)

Year	Staff strength	Staff Increase	Total sales 1000 quintals	Total Purchase 1000 quintals	Sales per staff in quintals	Purchase per staff in quintals
1980/81	1678		2161	1023	1288	610
1981/82	2040	21.57%	2213	1890	1085	926
1982/83	2734	34.02%	4801	2566	1756	938
1983/84	2825	3.33%	2246	3065	795	1085
1984/85	2946	0.74%	2161	3844	759	1351
1985/86	3065	7.70%	2213	2665	722	869
1986/87	3337	8.87%	4801	1148	1439	344
1987/88	4152	24.42%	2246	2721	541	655

Source: GSE Purchase and Sales showing per capita productivity in quintals

(c) Earning foreign exchange at the expense of local grain sufficiency raises the question of the justification of the strategy employed by the state in internationalizing grains and also whether the opportunity cost can be credited to synergy.

CONCLUSION

Since the end of the second world war, state-owned enterprises have expanded internationally either by export operations or by foreign investments. These enterprises have grown because the role of the state in economic affairs holds that governments have the responsibility for economic functioning of the national economy—providing industrialization and employment.

SOEs have been perceived as unfair competitors in the international market and their behaviors have been described as predatory in the international scene (Aharoni, 1980). In socializing grain distribution, Kongi State has decided to provide grain to the international market by charging only 50% of the cost of purchase and this is so done to provide a competitive edge to the product in the international marketplace. The implication is that the cost of generating foreign exchange will be borne by government in local currency and made good through subsidy, even though the strategy may be seen as unfair to other world producers. The social implications of this arrangement is that it discourages production motivation in grain business at the co-operative and farmer's level.

Grain socialization, whether for local or international distribution, must be given a sound base to motivate the producers through appropriate pricing and credits, otherwise the strategy would be shortlived and self-defeating.

REFERENCES

Aharoni, Y. (1980). "The State Owned Enterprises as a Competitor in International Markets," *Columbia Journal of World Business* (Spring).

Arpan J.S. (1972). "Multinational Firm Pricing in International Markets." *Sloan Management Review*, Vol. 14, pp. 1-9.

Day, A.J. (1976). *Exporting for Profit*. London, Graham and Trotman.

Kotler P. (1972). *Marketing Management: Analysis, Planning and Control*, 2nd ed. Englewood Cliffs N.J.: Prentice Hall.

Mackay D.I. (1964). "Exporters and Export Markets." *Scottish Journal of Political Economy*, Vol. II, pp. 205-217.

NEDO (1977). *International Price Competitiveness Non-Price Factors and Export Performance*. London: National Economic Dev. Office.

Parkinson, J.R. (1966). "The Progress of U K Exports." *Scottish Journal of Political Economy*, Vol. 13, pp. 5-26.

Piercy, N. (1982). *Export Strategy: Markets and Competition*. Boston: George Allen and Unwin, p. 10.

Robinson, R.D. (1967). *International Management*. New York: Holt Rinehart and Winston.

Terpstra, V. (1978). *International Marketing*, 2nd ed. Holt Saunders International Edition.

Tookey, D.A. (1969). "International Business and Political Geography." *British Journal of Marketing*, Autumn, pp. 136-151.

Chapter 12

Pricing Decisions in Global Marketing

Rita Martenson

BACKGROUND

In a recent survey of 303 executives by Fleming Associates it was clearly shown that pricing pressure was a leading concern for most companies today as well as the year before. Reasons for this concern could be found in the increased competition from worldwide markets, greater price awareness among consumers, and an increasing demand for high quality products. Price competition was found to be particularly important for older products, since price competition is a major tool to preserve market positions when product differentiation is harder to achieve, e.g., in the maturity life cycle phase. In another recent study, Boddewyn, Picard and Soehl (1985) found that approximately 90% of 71 American companies in the EEC thought that price competition in the EEC had increased over the past 10 years. No company thought that it had decreased. Approximately one third of these companies reported that they used price as a major competitive tool in the EEC. This result is hardly surprising, since the life cycle phases of new products have become shorter and shorter over the years (see, for example, Qualls et al., 1981). Thus, price decisions seem to be of utmost importance both nationally and internationally. However, if price decisions are difficult to make on the domestic market they will usually be even more difficult on foreign markets where several additional factors have to be considered.

PURPOSE OF THE STUDY

The purpose of this paper is to discuss pricing on international markets based on how a highly successful, global furniture retail firm (IKEA) priced its products on 8 different markets around the world. IKEA is most likely the largest furniture store in the world. In August 1985 it operated 18 stores in Scandinavia, 30 stores in the rest of Europe, including 17 in W. Germany, its largest single market. Outside mainland Europe IKEA also had 8 stores in Canada, 1 store in the U.S.A., and 12 stores in Kuwait, Australia, Hong Kong, Singapore, Canary Islands, Iceland and Saudi-Arabia. The stores mentioned vary in size from the large Stockholm furniture warehouse with 43,000 sq.m. to the smallest start boutique in Hong Kong with only 250 sq.m. Most European stores are in the range of 10-20,000 sq.m., however. IKEA furniture warehouses promote furniture under one single name worldwide. Based on a highly standardized core product, the company has achieved a clear global identity. The competitive strategy chosen by IKEA is a global cost leadership strategy. A global company achieves a strategic advantage from a coordination of its worldwide operations and will be willing to accept lower prices on some markets if this increases its worldwide profits. It is therefore quite likely that the competitive pressure on a certain market as well as that market's strategic importance for the global firm will influence its prices.

METHODOLOGY

The company chosen for this study was IKEA, a Swedish global furniture retailer. There are several advantages with this choice. First, by choosing a retail company it is possible to compare what the final buyer has to pay in different countries. This should be more interesting than to compare manufacturers' prices across countries. A manufacturer's price might be radically altered by more or less efficient resellers on the various markets. Second, the choice of IKEA made it possible to include a large number of countries, since the data collection was facilitated by IKEA's use of catalogs for its products. Totally 1,822 observations were made from IKEA's catalogs in 1980/81. The catalog is a very important promotional tool for the company and it could therefore be assumed that the products included in the catalog were representative for IKEA's assortment. The exchange rates used to convert the different currencies to Swedish Crowns were based on a yearly average for the period covered by the catalogs. Ideally, price comparisons should have been based on actual sales of different products in each country. This was not possible, however, and results presented here are thus unweighted. Third, there are very few retail firms operating on international markets, and few have been as successful as IKEA. In other words, IKEA belongs to a small group of unique retail firms.

The countries included differ in terms of size, distance from Sweden, competitive pressure, etc. and should therefore be a sufficient base for comparisons. The results achieved were also compared to earlier studies on IKEA's prices on some markets. In addition to the analysis of IKEA catalogs, several months were spent interviewing retailers in Sweden, W. Germany, Switzerland and Austria and reviewing secondary data on the different markets.

The first part of this paper is an integrated discussion of earlier studies and IKEA's operations on different markets, mainly Sweden, W. Germany, Switzerland and Austria. The second part of this paper is an analysis of IKEA's furniture prices on 8 markets.

THE ROLE OF THE PARENT COMPANY

Organization on World Markets

The competitive strategy chosen by the company must by necessity be reflected in the way the company is organized. If the company has chosen a global strategy, it must coordinate its operations worldwide to maximize global profits, instead of maximizing profits on individual markets. In addition, planning, budgeting, performance review and compensation must be done with the global profit in mind, which typically means that major decisions will be centralized. A global approach means that the company takes account of more than one country in its decision making. An example is that a certain market can be unprofitable per se, e.g., because of an on-going price war, but be profitable in a global perspective, since it enables the company to take advantage of economies of scale in production. If, on the other hand, the company had operated with a multinational approach, it would have looked at the contribution of each market per se. A market that wasn't profitable and showed no potential for becoming so in the near future would consequently be a good candidate for abandonment. The advantage with centralized decision-making is that it is easier to coordinate operations worldwide. The disadvantage is that it might be demotivating for subsidiary managers not to have full control of local decisions.

Strategic Markets

Some markets can be of strategic importance for the company, i.e., some markets can be more important than other markets. A market could be of strategic importance because of its size/potential or other reasons. California is, for example, considered a strategic market in the U.S. automobile industry, since many of the new trends start in California. Major automobile manufacturers want to be present on that market. The United States as a whole is considered a strategic market for many companies due to its large potential. The home market is usually also considered as a strategic market. No one wants to be a looser on the home market. Most companies will be willing to defend strategic markets if they feel threatened. Most companies will be willing to give additional support to subsidiaries operating on markets considered strategic. IKEA's strategic markets, W. Germany, France and the United States, were entered with much more care than the smaller markets, for example. German-speaking Switzerland was used as a platform for the entrance to the important German market; French-speaking Switzerland was similarly used as a platform for the more important French market, and Canada was the platform for the huge American market. Already in 1979, W. Germany had become IKEA's most important market in terms of sales. A large volume market can easily compensate for lower margins, which means that IKEA could "afford" lower prices on the German market and still get a large profit. IKEA's major European markets in relation to the firm's turnover there in 1979 were:

1. W.Germany	41%
2. Sweden	37%
3. Denmark	7%
4. Norway	4%
5. Switzerland	4%
6. Austria	4%
7. The Netherlands	3%
	100%

Nationality of the Headquarter

Businesses from different parts of the world are typically operated in different ways due to managerial traditions. European companies, and Scandinavian in particular, are typically allowed to operate much more independently than, for example, American businesses (Brandt and Hulbert, 1976, Hedlund 1984, Edström and Lorange 1986). Typically, the more freedom a subsidiary enjoys, the less standardized and coordinated the marketing mix. Uncoordinated freedom can create problems, however. Buzzell (1968) described a case where Philips Gloelampenfabrieken found that prices on some of its appliances in Holland were being undercut by as much as 30% by the company's German subsidiary. Costs were lower and competition more intensive in Germany. To this came the fact that German exports to Holland were encouraged by export subsidies from the German government. This is, of course, an undesired situation from the company's point of view, since it most likely creates bad will among customers as well as irritation between local subsidiaries.

Locus of Transfer Price Determination

Transfer pricing is the intracorporate system of international pricing, i.e., the price that one unit in the corporation sells products to another unit in the same corporation. The transfer price determines in which country the company is going to show most of the profit, and decision about transfer prices are therefore important decisions. Arpan (1986) consequently found that transfer prices were set at a high level in the company, probably because transfer prices can significantly alter financial results of global operations. Arpan also found that the greatest degree of participation by subsidiaries occurred when intracorporate transfers are infrequent but high in value, or when their operations are larger than their parents. Non-U.S. transfer pricing systems were found to be less complex and more market-oriented than American systems. U.S., French, British, and Japanese managements preferred cost oriented systems, while the Canadians, Italians and Scandinavians preferred a market orientation. Nationalities like the Germans, the Belgians, the Swiss and the Dutch did not exhibit any particularly distinctive orientation in preference. Arpan found it surprising that the Germans were among the least concerned with intracorporate pricing, since they exercised the closest control over subsidiary operations of any non-U.S. groups.

THE ROLE OF COMPETITION

Market Structure

The market structure in a country is how a market is built up of firms in different sizes, numbers and institutional types. The larger and more powerful the companies in a certain market are, the harder they will fight back when a foreign company enters. For example, the German market had very large companies which in addition cooperated in large and efficient joint buying groups. German retailers were consequently able to match IKEAs low prices to an extent not possible for retailers in Switzerland and Austria, where most retailers were small family businesses. However, the weak side was that retailers and manufacturers didn't cooperate. Manufacturers had been forced to manufacture a large number of different models, just because retailers demanded exclusive models to reduce the effects of price competition. German furniture manufacturers were therefore unable to offer competitive prices due to retailers' demands for exclusive models. Retailers had tried to compensate for this by having salesclerks work mainly on commission, but that created an unpleasant atmosphere in the store.

Entry Deterring Prices

When IKEA entered the German-speaking markets, the prevailing price level was perfect from IKEAs point of view; very high, which made IKEA's low prices even more attractive. IKEA could easily earn above-average profits from entry and the risks for retaliation were quite small.

Forms of Competition

It is well known that some forms of competition are less stable than others. Low prices and advertising claims are, for example, easily matched by competitors. When IKEA entered the Austrian market, there had been a constant price war between retailers, and the market could be described as turbulent. Retailers were therefore prepared for price competition at the same time as they were somewhat exhausted from the earlier price wars. It is too easy to match a price cut and the consequence is easily a reduced profitability for the whole retail community.

Competitive Position

In this paper competitive position means the position that the foreign company possesses in terms of familiarity, market share rank, etc. If the company is fairly unknown on foreign markets it could use a penetration pricing strategy to gain market shares, i.e., to set a price that is low enough for the majority on the market. However, if the company expects that competitors will imitate its offer, it might be wiser to set a skimming price in the beginning, i.e., to try to get as much as possible from the market before competitors have entered/imitated.

Globalization of Competition

On different markets and in different industries there will be various types of competition. Arpan (1986) found that the degree of competition in the host country market was almost always considered when companies made decisions about the transfer price. However, once a competitive position had been attained, many firms tried to maximize net world income by maneuvering profits to the lowest tax rate areas. In retailing it is different, mainly because of large cultural differences and the nearly complete absence of global customers. Retailing is typically characterized by companies with country-centered strategies. In other words, there are very few multinational and global companies in retailing. Very few retailers have to consider that their moves in one country can be linked to the moves of both competitors and its own company in other countries.

IKEA's Competitive Advantages

As a global competitor on domestic or multidomestic markets, IKEA had several competitive advantages that gave the company additional freedom in its pricing decisions. IKEA enjoyed, for example, large production economies of scale. With a standardized core product, IKEA could have several of the largest furniture factories producing the same item to supply the world demand. In contrast, IKEA's competitors, who had joined various buying groups, had difficulties cooperating on a much smaller scale, often because they could not agree on what specific furniture models to order. Only exceptionally would they get economies in scale in purchasing. In addition, IKEA did not own the furniture factories from which it purchased its furniture, and was therefore free to move its sourcing across the world. None of the buying groups had a chance to match IKEA's global experience and none was such a powerful buyer as IKEA. Instead, they selected a few of IKEA's best-selling models and sold imitations of these high-volume items at a lower price than IKEA's price. Most of IKEA's prices were fixed for one year. IKEA's global experience has also influenced its logistical system, which gives the company large economies of scale. For example, furniture is designed to reduce transportation costs, distribution centers are built to reduce transportation and storage costs, etc. IKEA also enjoys a competitive advantage because of economies of scale in marketing. Because of the rapid worldwide expansion IKEA has specialist teams that plan and open up new warehouses, and other marketing experiences from one market can easily be adapted to the other markets.

IKEA's Competitive Disadvantages

Differing product needs can act as an impediment to a global competition. The global retailer can thus have a reduced local competitiveness. Products with strong roots in local tradition typically don't have an immediate world demand, e.g., food and furniture. As a global retailer IKEA had to take the risk that the products it sold would not be accepted by other cultures. Experts in the furniture industries in Sweden, Germany, Switzerland and Austria, typically predicted that there wouldn't be a market for the type of furniture that IKEA sells. Furniture design in the German-speaking countries was quite different from the IKEA-design. The German-speaking countries were used to much larger and darker furniture than the light pinewood furniture offered by IKEA. However, when the price difference is large enough, there will always be enough people who are willing to try something different. A problem created by IKEA's size is that planning can be difficult. If some products become instant successes on all markets, they will probably be out of stock and create many disappointed customers who came to buy what was presented in the catalog. Traditional furniture retailers don't have the same expectations to live up to, since customers know that they have to wait a few months for their new furniture bought at a traditional store. IKEA's customers thus have a higher sensitivity to lead-times due to expectations created by IKEA.

THE ROLE OF THE PRODUCT

It has already been mentioned that German furniture retailers, for example, demanded exclusive models as a shelter against price competition. Product differentiation as well as the exclusive right to sell a certain product is thus something desirable, since it creates stability on the market. Switching costs for the buyer have the same stabilizing effect. Switching costs are one-time costs facing the buyer for switching from one supplier to another. Switching costs can be employee retraining costs, costs for spare parts that no longer fit, costs of new ancillary equipment, etc. When IKEA entered the U.S. market it didn't change the measures on its beds, which are different from the American standard. Since Americans are used to fitted sheets, buyers of IKEA-beds would consequently have to pay for switching to IKEA's bed-sizes. The advantage, of course, was that once buyers had been pursuaded to buy IKEA's sizes, they would encounter switching costs for new sheets if they changed back to the American standard. However, with a low price on the beds, IKEA could recapture additional profits from selling more sheets, etc.

Product differentiation has become harder to achieve over the years as more companies offer more and more products that are more or less identical from the buyer's point of view. Consequently, it is often no longer possible to use physical or technical differences as a basis for product differentiation. Product differentiation created by advertising, i.e., the image a product is given, is harder to achieve and is easier to match by competitors. In several cases the market leader has had such a tremendous advantage from having established a strong brand franchise in a time when that was easier, that the market leaders' position has been impossible to threaten.

If the product is standard or undifferentiated buyers can always find an alternative supplier. An excellent case is IBM's personal computers. Despite IBM's brand identification and earlier customer loyalties, it turned out that the price difference between IBM's original products and the compatible PCs was too large. IBM lost half the market to other companies, despite the fact that IBM was first on the market.

The basis of buyers' choice can sometimes be broadened. When IKEA entered the Swiss market in 1973, Swiss retailers had concluded that there would be no Swiss buyers who would be willing to carry furniture of normal size under their arms or to bring them back home in their car. Cash-and-carry was consequently not seen as a realistic alternative in Swiss furniture retailing in the early 1970s. IKEA's success showed that it was possible to broaden buyers' choice alternatives.

Even if the physical product is the same on all markets, it will not be perceived in the same way on all markets. Car manufacturers like Volvo and Saab have made the Swedes used to their automobiles, whereas Renault, Peugeot, and Citroen have made the French expect certain things from a car based on their experiences with

Renault, Peugeot and Citroen. Consequently, buyers will perceive products based on what they are accustomed to expect from similar products. German, Swiss, and Austrian retailers were highly annoyed that IKEA created the "wrong" kinds of price expectations. Having seen IKEA's prices, many consumers started to question whether they would really have to pay two or three times more from another retailer. Retailers in the German-speaking countries thought that they had to spend too much time explaining why furniture in different materials, e.g., wood, has different prices. Similarly, many Swedish industrial sellers operating on the U.S. market think that they have a large disadvantage from selling products with Swedish standards. The National Swedish Board of Occupational Safety and Health has, for example, developed standards that make products (e.g., trucks) much more expensive to produce. Since American buyers have other priorities, they are often not willing to pay extra for things designed for worker's health. The fact that it is both foreign and different from their own expectations about the product creates a resistance.

THE ROLE OF DISTRIBUTION CHANNELS

Channels of Distribution and Distributor Margins

Even if the manufacturer wants to have the same price across markets, this could be difficult to achieve because of differences in reseller efficiency and power between markets. A study of cost variability for medicines in Germany, Switzerland, France, U.K., and Italy, showed, for example, that wholesalers' markup varied from 10 to 25 (Italy respectively Germany). Retailers' mark-up varied from 38 to 99 (Italy respectively Switzerland). Similarly, VAT (value added tax) varied from 0 to 34 (Switzerland respectively France), which in the end showed that manufacturers' initial prices of 100 had increased to 157 in Italy but to 242 in Germany (Terpstra, 1983).

If the company is small, new and foreign, it must be prepared to get the least attractive channels of distribution. There is no reason why the best manufacturers' representatives would take the chance with a small, foreign newcomer with an uncertain future, if they can work for a large, stable, and profitable national firm instead. Existing competitors may even refuse to deliver to channels which accept the foreign newcomer. Sometimes these barriers to entry are so high that to surmount them the new firm must create an entirely new distribution channel. Since IKEA is a retailer, there was no such problem. However, a strong relationship between manufacturers and retailers created lots of similar problems for IKEA in Sweden in the mid-sixties. Furniture manufacturers who were forced not to deliver to IKEA because of retailer pressures were a major reason why IKEA went to Eastern European countries for sourcing.

THE ROLE OF CUSTOMERS

Customer Loyalty

Nothing is more profitable than a loyal customer. Therefore, the company that has a large stock of loyal and satisfied customers will have much lower costs for communications and administrations than the company that has few loyal customers. It is so much cheaper just to remind the customer about your product, than to persuade buyers of competing brands to switch.

Future Potential

The future potential of the customer should have a major impact on the company's pricing strategy. A lower price could be accepted if the customer is expected to be profitable in the long run. One such reason could be that the customer is small now but has a very positive growth potential. The buyer's growth potential could be determined from the growth rate of the industry in which it operates. It could also be determined by the growth rate of its primary market segments.

The growth potential of households or consumers depends on both socio-economic factors and cultural factors. In Switzerland, a higher income meant a higher consumption of furniture, absolutely and relatively. In W. Germany, furniture consumption in DEM increased with increasing income, but relatively, less was spent on furniture in the higher income brackets. In Austria, by tradition, very little was spent on furniture and on the home as such. Austrians had a much lower standard of living than the Scandinavians, the Germans and the Swiss. Austrians wanted to spend as little as possible on furniture, so that they could spend as much as possible on other things. The price sensitivity was consequently positive for IKEA, whereas the lack of interest in furniture was not. Generally, however, buyers earning low profits or having low disposable income are not a profitable segment to work with, because their incentives to lower purchasing costs will make them too focused on price alone. Another thing that influences the consumption of furniture is the ownership of summer houses. Swedes probably have more summer houses than other nationalities, which means that the potential for selling furniture is increased.

When the costs of the product are a small part of the total costs and if the product is a relatively low-cost item, then the perceived benefits of price shopping and bargaining tend to be low. An example from furniture retailing is all the home-furnishing items sold by IKEA; cooking-ware, pots, textiles, etc. Despite the fact that each item costs very little, the volume and the margins make this sector a highly profitable one. Once customers have come to IKEA it is convenient to purchase these items there and it is much harder to compare these prices than to compare the price on different high-price items like sofas.

Service Demanded and Cost of Serving Buyers

In Germany, Switzerland and Austria, retailers generally included delivery and assembly in their prices and they therefore used a higher markup on wall units and book shelves than on sofas, for example. The former were much more difficult to assemble, they often had to be customized and were easily damaged. IKEA used a quite different business philosophy. With a completely standardized core product, it offered all customers within a country the same prices. To help customers needing more service, IKEA instead established agreements with outside firms to provide such service (e.g., small truck rental).

Price Sensitivity

The more important each purchase, the more price sensitive the buyers could be expected to be. The importance of a purchase depends both on the volume involved and in what period it is purchased. When IKEA entered the German market in 1974, the market was really depressed and the upsurge lingered until 1976. One of IKEA's managers thought that this was a great advantage for IKEA, since depressed periods would make consumers even more price conscious and even more prepared for changes and different alternatives, particularly if these changes are associated with price advantages. This conclusion was true in the case of IKEA and it was even more so during the depression of the 1930s in the U.S., when the low price institution of its time had a great success. In addition to consumers being more willing to accept new offerings, competitors would be more reluctant to take risks and invest. Passive competitors mean lower entrance costs.

Comparability

The easier it is to compare prices, the larger the role of pricing. IKEA has an unusually high visibility of its prices because of the comprehensive product catalog. The catalog is an excellent promotional tool, but it can create certain problems because of its lack of flexibility. Since the catalog is available to competitors as well, it is easy for them to reduce their prices on certain items knowing that IKEA's prices are fixed for one year. Similarly, if IKEA is forced to pay much more than expected for certain products, the company might be forced to sell some products at a lower price than it pays itself until the new catalog arrives.

Discounts

German retailers had offered customers such large discounts that it was sometimes said that customers were more interested in buying discounts than merchandise. For IKEA this was no alternative, since IKEAs business definition was based on low prices that were fixed for one year at a time.

THE LIFE CYCLE PHASE

If the company sells unique or differentiated products it could choose a creaming price strategy in the introduction phase. A creaming price strategy is based on high prices and high margins. A creaming price strategy could be used if the company assumes that there are certain buyers who are willing to pay more for being the first owners of the product. A creaming price strategy could also be motivated if the company knows that it has a small competitive advantage, but that it will only be a question of time how long this advantage will last. This is the risk with a creaming strategy, i.e., that it attracts new entrants to the market. The opposite to the creaming price strategy is a penetration price strategy. A penetration strategy is based on volume and lower margins. For IKEA, a mass retailer, this was the only alternative available. IKEA needed an instant success to pay back the enormous investments made, not the least in the launch campaigns. Another reason for IKEA to choose a penetration strategy was its business idea; the business idea assumes that customers visit IKEA a couple of times per year and if they don't make major purchases of furniture each time there is always smaller things for them to buy. Later in the life cycle most companies do not have as many alternatives as in the introduction and growth phase. With more competitors, there is more price pressure. Noncompetitive companies are squeezed out of the market, which results in increased stability of market shares. On a stable market there is also more stability of the price structure; price wars reduces profitability for all companies.

THE ROLE OF GOVERNMENT

Dumping

Sometimes it might be wise to charge a uniform price on all markets just to avoid charges of dumping in the host country. In the U.S. it is well known that foreign companies often sidestep antidumping restrictions by taking their products over countries like

Canada. In Europe there are other countries over which merchandise is shipped to avoid import restrictions, etc.

Price Differentiation

If prices vary too much between countries it might attract undesired attention. Davidson (1982) described, for example, how official inquiries into Hoffman-LaRoche's pricing for Valium and Librium came as a result of information showing that prices for these products varied up to 400% from market to market.

Consumer Interest Organizations

The West German consumer association AGV (Arbeitsgemeinschaft der Verbraucher) had accused the furniture retail trade of having excessive profit margins. According to AGV, furniture retailers acted essentially as showrooms for manufacturers. Retailers in general did not stock furniture but placed only the order once the goods had been sold to the consumer. Consumers thus financed part of the retailers' business with their deposits.

INTERNATIONAL PRICE SETTING ALTERNATIVES

Cost Approach

A cost approach to pricing means that the focus of the company is set on the costs. Based on costs a profit margin is then added to arrive at the final price. This is a very simple approach and is commonly used in retailing. Retailers, and in particular big item retailers like furniture retailers, have been highly criticized for using this approach, since it has been common to use the same markup on high price items as on low price items. It thereby becomes more profitable for the retailer to sell an expensive sofa than to sell an inexpensive sofa. Another factor is that there are usually differences in country costs. Whether or not this will play any role depends on the company in question. A company that operates on one market only, or with a multinational approach where each market has to be profitable, would have no choice but to accept the cost difference.

For a global company it would be different. A global company operates to maximize the global profit. This might imply that it could be profitable to accept high costs in one market, because this market enables the company to take advantage of the economies of scale in production. Another reason for accepting unusually high costs could be that the company has global sourcing.

Market Approach

A market approach to pricing means that the company estimates how much the market can bear. If the market can accept a high price it is good, otherwise the company will have to consider whether or not that particular market is important enough to accept a lower price than wanted. In terms of transfer prices, Arpan (1986) found that the normal Scandinavian pattern is to use market prices when they exist, and to approximate them when they do not.

Export Pricing

Export pricing is a special situation, since new kinds of costs have to be added like transportation costs, custom duty, insurance, etc. Not only are these costs additional, but they will vary from one country to another.

COORDINATION OF PRICES ON INTERNATIONAL MARKETS

Compared to product decisions, pricing decisions are typically much more dissimilar from country to country. One reason for price changes as well as price differences are the heavily fluctuating exchange rates. For example, in the period 1980-1984 the exchange rate for the U.S. dollar varied from 4.21 Swedish crowns to more than 9 Swedish crowns. Whatever price policy the company has chosen, the tremendous size of these fluctuations will make most price decisions a difficult matter. Whether or not final customers will notice depends on the pricing policy chosen. Keegan (1974) has suggested three major alternative pricing policies in international marketing:

1. *an ethnocentric pricing policy*, where the price of an item is the same all over the world and customers absorb freight and import duties
2. *a polycentric pricing policy*, which allows the subsidiaries or affiliates to establish whatever price policy they consider to be most appropriate on their local market
3. *a geocentric pricing policy*, which takes an intermediate position by neither fixing a single price worldwide nor making it a completely local decision.

The pricing policy chosen will be influenced by the extent to which the decision making is centralized or decentralized in the company. An ethnocentric pricing policy requires centralized decision-making, whereas the opposite is true for a polycentric pricing policy. Boddewyn and Hansen (1977) as well as Boddewyn, Picard and Soehl (1985) found that pricing decisions were typically not a local decision. Most companies in these two studies seemed to make price decisions at the headquarters and the second most common alternative was the joint decisions. Producers of consumer durables had increased the number of price decisions made jointly and producers of consumer nondurables had increased the number of price decisions made at headquarters (based on the assumption that the samples were comparable). Aylmer (1970) found that the retail price was a local decision for the large majority of companies (74%).

It could be an advantage to use an ethnocentric pricing policy in case the company has global customers and is selling a product with a low degree of differentiation. In such a case customers would easily be able to compare prices from one market to another. Global customers are usually very large customers and can easily switch from one market to another if they get a price advantage from such a shift. In addition, it would create too much badwill if prices for identical products varied too much between markets.

The reasons for price adaptations between countries can be of two types (Sorenson and Wiechmann, 1975). Obligatory custom-tailoring are adaptations that the company is forced to make, maybe due to the legal situation in a particular country. Discretionary custom-tailoring are adaptations that the company chooses to make. It might sometimes be impossible to coordinate prices between markets because local needs and buying behavior might vary extensively from one market to another. These researchers found nevertheless that the majority of the multinational companies (56%) included in their study had a high degree of retail price standardization, 14% had a moderate degree of standardization and 30% had a low degree of standardization. Leksell, Spangberg and Lindgren (1981) who studied Swedish firms on the U.S. market found for example that the U.S. purchasing departments were unwilling to pay any premium prices because they had bonus systems based on the difference between standard costs (according to the specifications) and the final price.

A fairly new study by Arpan (1986) showed that Scandinavian firms operated somewhat differently from other firms. Scandinavian firms normally used market prices when such existed and approximated them when they did not exist. With their small home markets, Scandinavian firms considered host-country acceptability a major concern.

RESULTS

An analysis of IKEAs prices shows that IKEA has a low price image on all markets. Nevertheless, when the company entered the Austrian market something quite unexpected happened: IKEA was accused of charging too much! Based on an Austrian consumer interest organization's investigation of IKEA's prices, Austria's trade minister, Herrn Staribacher, estimated that prices in Sweden were 60-70% lower than the Austrian prices. The study included IKEA's catalog prices in Denmark, Germany, Switzerland and Austria in 1978.

Table 12.1 shows that IKEA's prices on the Swiss market are much higher than for example on the German market. If the VAT, similar to sales tax, is considered the differences are even larger. It is questionable however whether the conclusion drawn by the Austrian consumer interest organization was correct. The analysis clearly shows that IKEA's products were more expensive in Austria than in Denmark. However, when the VAT is considered one finds that prices were lower in Austria than in Switzerland, despite the fact that IKEA had lots of problems with deliveries to Austria, with their outlet, etc. Analysis of prices in the eight countries included here confirms the earlier results, see Table 12.2. The price level in Germany is much lower than the price level in Switzerland and Austria. Considering the geographical distance, prices in Canada are also quite low.

How can this result be explained? One reason for the higher prices on the Austrian market was that IKEA didn't own its showrooms in the outskirts of Vienna and rents made it more expensive to operate this store. Another major reason was the competitive

TABLE 12.1. IKEA's Prices in Denmark, Germany, Switzerland and Austria 1978 (309 items)

Country	Denmark	Germany	Switzerland	Austria
Prices incl. VAT - Index	100	118	128	134
VAT (Value-Added-Tax, similar to sales tax)	18%	12%	5.6%	18%
Prices excluding VAT - Index	100	124	143	134

TABLE 12.2. IKEA's Prices in Eight Countries 1980/81 (1,822 observations)

Country	Index*	Minimum Value**	Maximum Value**
Sweden	100		
Denmark	113	76	146
Norway	119	38	161
Germany	120	72	158
Switzerland	134	84	187
Austria	140	67	234
Canada	139	62	269
Australia	180	114	309

* Average values

** The minimum and maximum values show how much cheaper or more expensive identical products are compared to the Swedish prices.

pressure or lack of such a pressure. The Austrian as well as the Swiss retailers were very small and did not have enough resources to compete with the global IKEA. Neither Switzerland nor Austria were large enough to be considered strategic markets. The German market was the most important foreign market for IKEA. The German retailers were, on the other hand, large and used to aggressive competition. Some of them even imitated IKEA's furniture and sold these products cheaper than IKEA. On the German market there were also large joint buying groups which coordinated their purchases to get more favorable purchase prices. Furniture manufacturers in Switzerland had in some cases been so upset with the inefficient Swiss retailers that they had opened their own showrooms. Retailers interviewed in Sweden, Switzerland, Germany and Austria thought that IKEA was able to have low prices and high markups because of the favorable purchase prices the company achieved from factories in Eastern Europe, for example. This result also points at the large benefits a global company can achieve compared to companies operating on one market only; by charging more on less competitive markets it can "afford" to charge less on more competitive markets and thereby achieve a market position that is superior to what could have been achieved by the local companies. Another factor that can explain price differences is that it is generally less expensive to supply a larger market than many smaller markets. IKEA opened distribution centers on the German market, which because of its size was the most important market for the company. It could be assumed that the Australian market was less competitive because of its geographical location combined with the fact that its size makes it less attractive than the German market, for example. The German market is more than three times as large and is probably much easier to enter considering both geographical and cultural differences. In addition, IKEA's operations on the Australian and Canadian markets differed from its operations on the other markets, in that the former were based on franchising. Later IKEA took over the Canadian franchise operations itself, since Canada is an important market to have as a base for entrance to the large U.S. market.

There is another quite interesting aspect of IKEA's pricing policy. Prices in its catalogs are fixed for one year. When there is high inflation and exchange rates fluctuate largely, this is a risky business but most likely highly appreciated by consumers. A recent analysis by Forbes magazine showed that many well known foreign firms on the U.S. market changed prices much less than the fluctuations in exchange rates, but there were nevertheless quite visible changes in final prices. It is quite likely that it is more important to use IKEA's low prices in the catalog as a promotional tool, than to be able to change prices when changes are justified. It is also most likely that IKEA's suppliers can't change their prices either, which reduces the risk IKEA takes. Another quite interesting aspect of IKEA's pricing policy is that margins are quite often higher on less expensive items and lower on more expensive items. This fact is, of course, a limitation with the conclusions drawn in this study, where it wasn't possible to consider how much IKEA sold of different items in the different countries. The overall conclusion is, however, not affected by this limitation, i.e., that IKEA operates as a truly global retailer and benefits from the lack of competitive pressure on some markets to make it stronger for intense competition on other markets.

REFERENCES

Arpan, J.S. (1986). "International Intracorporate Pricing: Non-American Systems and Views." In Dymsza, W.A. and Vambery, R.G. (eds.) *International business knowledge. Managing international functions in the 1990s*, New York: Praeger.

Aylmer, R.J. (1970). "Who makes marketing decisions in the multinational firm." *Journal of Marketing*, October.

Boddewyn, J.J. & Hansen, D.M. (1977). American marketing in the European Common market, 1963-1973. *European Journal of Marketing*, Vol. 11, No. 7.

Boddewyn, J.J., Picard, J. & Soehl, R. (1985). "U.S. Marketing in the European Common Market, 1963-1983: A Longitudinal Study." The Bernard M. Baruch College of the City University of New York, New York.

Brandt, W.K. & Hulbert, J.M. (1976). "Communications and Control in the Multinational Enterprise." In *Multinational product management*, Proceedings, American Marketing Association/Marketing Science Institute Research Workshop, August, Report No 76-110.

Buzzell, R.D. (1968). "Can You Standardize Multinational Marketing? *Harvard Business Review*, November/December.

Davidson, W.H. (1982). *Global Strategic Management*. John Wiley & Sons.

Edström, A. & Lorange, P. (1986). "Matching Strategy and Human Resources in Multinational Corporations. In Dymsza, W.A. & Vambery, R.G. (eds.) *International business knowledge. Managing international functions in the 1990s*. NY: Praeger.
Forbes, May 18, 1987.
Forbes, June 29, 1987.
Hedlund, G. (1984). Organization In-Between: The Evolution of the Mother-Daughter Structure of Managing Foreign Subsidiaries in Swedish Multinationals. *Journal of International Business Studies*, Fall.
Keegan, W.J. (1974). *Multinational Marketing Management*. Prentice-Hall, Inc.
Leksell, L., Spangberg, K. & Lindgren, U. (1981). *Market Entry and Growth Strategies in the U.S. – the Swedish Experience*. Svenska Handelsbanken International Seminar, Stockholm, Sweden.
Martenson, R. (1981). *Innovations in Multinational Retailing*. Diss., University of Gothenburg, Sweden.
Martenson, R. (1987). *Swedish Marketing in the U.S.* Studentlitteratur & Chartwell-Bratt, Sweden and England.
Porter, M.E. (1980). *Competitive Strategy*. NY: The Free Press.
Qualls, W., Olshavsky, R.W. & Michaels, R.E. (1981). Shortening the PLC—An Empirical Test. *Journal of Marketing*, Fall.
Segmentation Strategies Create New Pressure Among Marketers. *Marketing News*, Fleming Associates, March 28, 1986.
Sorenson, R.Z. & Wiechmann, U.E. (1975). How Multinationals View Marketing Standardization. *Harvard Business Review*, May/June.
Terpstra, V.B. (1983). *International Marketing*. The Dryden Press.

Chapter 13

The Planned Development of Channels of Distribution in a Highly Centralized Socialist System – The Romanian Experience

Jacob Naor

THE ROMANIAN CONTEXT

It is only fairly recently that Romanian authorities have shown interest in marketing. The traditional focus of all socialist countries, and indeed of developing countries in general, has been on production and industrial growth. In Romania rapid economic growth has been a cornerstone of governmental development policy and remains so to this day. Romania can thus boast of having achieved some of the highest industrial growth rates in the world (Tsantis, 1979, p. 5). And economic targets for 1981-1985 for example, while lower than those of previous years, were still claimed to be amongst the highest in the World (Romania – Socio Economic Developments, Bucharest, Romanian News Agency, 1982, p. 37). It is not surprising therefore that a strongly entrenched production orientation continues to guide Romanian policy decision making. Yet changes have appeared of late. Thus, the Basic Home-Trade Law of 1972 lays the ground rules for marketing activities, stating that such activities were instrumental in raising the living standards of the population (1972, p. 1).

Romanian economic development, in no small measure a result of the aforementioned rapid-growth policy, has been traditionally characterized by sellers' market conditions, a highly monopolistic production and distribution structure, and a highly centralized and bureaucratized comprehensive planning system. These features continue to characterize the system to the present.

The highly monopolized Romanian production and distribution system, still in place to date, must be highlighted as of major relevance to the discussion at hand. It is totally in line with traditional command-type central planning of the "top-down bottom-up" variety, which has probably been maintained longer in Romania than anywhere else in the East-bloc. However, even there changes have been occurring, particularly since 1978 (Naor, 1982). Some of these changes as they pertain to marketing will be pointed out in subsequent sections.

Lastly, Romania had traditionally adhered to orthodox socialist-Marxist doctrines. Such doctrines have clearly affected and continue to affect the character of distribution in Romania. In this Romania is not alone. These doctrines, to one degree or another, have shaped the character of economic activities of all East-bloc countries and may be expected to continue to do so.

SOME PRINCIPLES AND PLANNING METHODOLOGIES UNDERLYING DISTRIBUTION

From available sources the following appears to underlie distribution in Romania: it is based on the centralized planning of networks, aims at achieving equitable distribution in a cost effective manner, and appears to be designed to meet "rational" consumer needs. These principles are similar to the principles underlying marketing in general elaborated elsewhere (Naor, 1984). While never stated explicitly in this fashion, they appear to underlie the Romanian Home-Trade Law, (1972) as well as various subsequent official pronouncements on marketing subjects.

The planning of trade networks and the administration of trade in general occurs within the framework of the overall social-economic national development plans (annual and five-year plans). It is entrusted to the Ministry of Home Trade which collaborates to that effect with other centralized bodies, the Central Union of Consumers' Cooperatives and the various administrative district councils. It is thus a centrally planned activity implemented and administered jointly (through appropriate checks and approval requirements) by the Home-Trade Ministry and related regional, "social" or state organizations. Table 13.1 presents a simplified outline of the methodology for planning the production and distribution of consumer goods.

As indicated, the planning of industrial production needed to meet consumer demand starts with the determination of such demand. This is done for both the five-year plans (in less detail) and their component annual plans (in greater detail). Plans, as indicated, are worked out by central organs in collaboration with regional bodies. Various planning norms are used in the process, the major ones being commodity/service per capita consumption norms which are prepared for major types of goods and then aggregated to provide total (approximate) industry demand.

While earlier pronouncements, such as those contained in the Home-Trade Law (1972, p. 2) stressed that existing consumer needs and preferences were to be carefully ascertained (and presumably used for norm setting purposes), subsequent pronouncements stressed that "rational consumption," to be scientifically determined by central research organizations in collaboration with industry and trade organs, should determine the setting of consumption norms (Fazekas, p. 52). Work on the determination of such "scientific" norms appears to be in progress (Ibid, p. 52). The Romanian authorities appear to be using both methods concurrently at the

TABLE 13.1. Simplified Planning Methodology for the Production and Distribution of Consumer Goods*

Planning Domain	Major planning Norm
Consumption requirements	consumption norms
Retail facilities needed	utilization norms
Wholesale capacity needed	capacity norms
Output needed	volume and quality norms

*Based on information supplied to the author in personal interviews.

present, presumably with the hope of being able to move increasingly towards a reliance on scientific norms in the future.

Having determined final demand, the need for retail facilities can be determined, based on appropriate industry specific retail facilities utilization norms. Such norms are issued by the Ministry of Home Trade for retail trade under its jurisdiction (Home Trade Law, p. 15). They presumably would deal with size and number of facilities by type of outlet, staffing requirements, and facility design. The planning of wholesaling capacity needed to service the retail network would then follow. Needed capacity norms, including warehousing and transportation norms would again be issued by the Ministry of Home Trade. As a general rule the authorities appeared to utilize a single or two wholesaling "enterprises" per administrative region of the country (Judet), of which there were forty.

The second distribution principle, equity, is based on the Socialist principle of "to each according to his contribution," based on the amount and quality of work performed (Fazekas, p. 64), with differentials constrained to a 1:6 ratio between the lowest and highest wage (Ceausescu, 1981, p. 508). Thus consumption norms, particularly those to be scientifically determined, were intended to reflect such work efforts, as well as other pertinent environmental conditions (Ceausescu, 1978, p. 10). In distribution, equity was implemented through the use of standard designs of retail facilities by industry. Such standards took account of urban/non-urban differentiation of facilities, based primarily on size. While this tended to affect assortments and line-depth, the official policy of attempting to equalize living standards nation-wide continues nevertheless to be pursued, and considerable progress has already been made toward reducing differences in consumption per capita between the different territories of the country (Fazekas, p. 69).

Cost efficient distribution, the third principle, was consistently stressed by the authorities, increasingly so since the early 1970s when the need to save dwindling domestic raw material resources was first recognized. The emphasis in distribution centered on reduction of transportation and warehousing costs. The ongoing urbanization facilitated considerably the shortening of distribution networks. Concurrently, government policy mandated the location of production and warehousing facilities in close proximity to areas of concentrated demand such as major towns and population centers.

Thus the principle of using direct distribution, to the extent possible, was promulgated and actively implemented (Ceausescu, 1978, p. 16). Production was to be increasingly synchronized with consumption to avoid excessive inventory buildups. Where feasible, goods were to be delivered directly from producers to retail outlets, or in the case of consumer durables, from wholesale warehouses to final consumers (Fazekas, 1978, p. 57). As will be seen later, this policy saw vigorous implementation.

It appears reasonable to presume that the authorities were aiming to provide some predetermined service levels, embodied in the planning norms used. Such levels presumably constrained the efficiency drive pursued by the authorities. Nothing is known however as to how conflicts, which undoubtedly arose between those aims, were resolved, if at all. Thus, it is not clear if, and to what extent, the shifting of functions and costs to consumers, which presumably occurred in the case of direct distribution, were included in the authorities' calculations. It appears safe to presume that the major focus of efficiency efforts would most likely be centered on the "visible" portions of costs, i.e., those costs incurred by the state, or cooperative trade organizations.

ORGANIZATIONAL COMPONENTS OF THE SYSTEM

Table 13.2 presents an overview of some major organizational aspects of the distribution system. Noteworthy is the almost exclusive role of cooperative organizations in handling retail trade in rural areas, which appears to be based on a tradition pre-dating World War II. Of the three basic existing forms of cooperatives, consumer, agricultural, and handicraft-artisans, the first was by far the largest, handling approximately 20% of total trade turnover at retail. The other forms handled the remaining 3% of trade turnover. All cooperatives handle production, as well as distribution and the provision of services. The artisans cooperative, for example, provides services primarily in the urban areas related to the goods they produced. The consumer cooperatives handled, amongst others, the

TABLE 13.2. Selected Organizational Aspects of Distribution*

General	Percent of trade turnover	
There are 40 regions (judets) in Romania. Trade is handled in primarily rural areas by cooperatives**, in urban areas primarily by state organizations. All wholesalers are organized as "state enterprises" and are under Ministry of Home Trade.	cooperatives	23
	state organizations	77

Wholesalers (selected)		Percent of distributed volume handled***
Food:	1/region, handles State enterprises and cooperatives	40
Textile:	18 total, 1 per 2-3 regions	80-90
Hardware:	17 total, 1 per 2-3 regions	60-70

Retail (State only-selected)		Percent of total State Trade
General merchandize - under Home Trade Ministry		80
Ministry outlets:		
Pharmacies	Health ministry	
Gasoline distribution	Chemical Industry Ministry	20
Fruits & Vegetables	Ministry of Agriculture	
Other	Other	

Retail outlets maintained by 10 economic ministries as well as by some Centrals (Headquarter-type organizations) and large individual enterprises.

Channel Markups	
Total markup:	13% ****
Wholesale markup	4.3 % (1/3 of total markup)
Retail markup	8.7 % (2/3 of total markup)

* Information provided to author in personal interviews by a former deputy minister of Home Trade.
** More than 60% of total agricultural output is produced by cooperatives.
*** Until 3 years ago. Current percentages are lower. Remaining volume physically by-passes wholesalers.
**** Computed on the basis of production costs.

procurement of agricultural goods from (private) households and their sale in both urban and rural areas. The number of cooperative retail outlets exceeded 38,000 (or 48.2% of total retail outlets) by late 1981 (see Table 13.3).

The highly monopolistic character of wholesale distribution is apparent from Table 13.2. The source of the information, a former deputy minister of Home Trade, assured the author that no thought was given to the possibility of increasing the number of wholesalers, a move that could achieve efficiency through competitive

TABLE 13.3. Number of Retail Outlets by Type of Sector

	Sector				
	Socialist			Private	Total (all sectors)
	State	Cooperative	Total		
1948	1952	8459	10411	97946	108357
1950	7762	13805	27567	38228	65795
1965	24550	28821	53371	-----	53371
1975	34966	36648	71614	-----	71614
1980	40582	38164 *	78746	-----	78746
1981	40873	38093	78966	-----	78966

*of which 33836 units, or 88.6%, were under the jurisdiction of the consumer cooperative movement. Statistical Yearbook of the Socialist Republic of Romania, Bucharest, Romania 1982, p. 240.

pressures. Efficiency, it was unequivocally stated, would be achieved through "the plan" i.e., through the setting of minimum performance standards in the plan that must be achieved to avoid monetary penalties, while rewarding performance that exceeded such standards incrementally. Planned monetary incentives, it was hoped, rather than unplanned uncontrolled incentives resulting from competition, would thus bring about efficiency in distribution.

As indicated the Home Trade ministry in collaboration with regional administrative councils had jurisdiction over much of retail trade as well as over all wholesaling activities. Starting with the economic reforms of 1978, territorial organizations had acquired considerable autonomy in trade matters not previously granted them. They now staffed, managed and controlled trade activities in their respective territories (Fazekas, p. 50). Operational decision making was thus brought closer to the operational level. The basic planning and resource allocation decisions remained centralized at higher levels, however. In addition, individual ministry outlets handled about 20% of total state trade turnover, a significant portion indeed. Considerations of efficiency and expertise were mentioned as the main reasons for this curious organizational arrangement.

As stated earlier, efficiency considerations were behind the drive to bypass wholesalers, which appears to have seen rapid implementation. For example, in the case of the largest Textile and Footwear wholesaling "enterprise" in Romania, the percentage of goods previously warehoused, and now shipped directly from manufacturer to retailers (for which the enterprise continued to act as a broker) rose from 6% of total enterprise turnover in 1978 to over 40% by late 1982 and was expected to rise further.* Shorter distribution networks were apparently being introduced rapidly indeed.

THE PLANNED DEVELOPMENT OF RETAILING

Some important trends in the development of retailing indicating the planned nature of distribution are evident from Table 13.3, which presents the number of retail outlets since 1948 by type of sector (state, cooperative, and until the early 1950s, the private sector). Notable is the continuing growth in the state sector since 1975, compared to the relative stability in the number of cooperative outlets since that time. This is in line with the ongoing trend of urbanization in Romania, and the fact that urban distribution was handled by the state. Since the state sector handled over three times the trade volume handled by the cooperative sector (77% as against 23%) and given the rough equivalence in the number of stores in both sectors, state stores must have been considerably larger than cooperative outlets. This is supported by data in Table 13.4, which provide the changes in the number of employees engaged in retailing since

TABLE 13.4. Personnel Employed in Retail Distribution (Socialist Sector)

	State		Cooperative			Food related distribution included in total:
	Total	Employees per Outlet*	Total	Employees per Outlet*	Socialist Sector Total	(Socialist Sector)
1950	82,100	10.6	59,600	4.3	141,700	33,100
1965	203,900	8.3	88,900	3.1	292,800	79,600
1975	304,100	8.7	91,000	2.5	395,100	142,800
1980	353,300	8.7	94,700	2.5	448,000	162,000
1981	355,800	8.7	94,900	2.5	450,700	165,600

Statistical yearbook of the Socialist Republic of Romania, Bucharest, Romania, 1982, p. 246.

*Computed by the author, using the number of outlets provided in Table 3.

*Based on information collected by the author in personal interviews.

1950. State outlets, on the average, employed three times as many employees as did cooperative outlets.

The planned nature of distribution is apparent as well from the stability in employees per outlet for the two categories of outlets, particularly since 1975 (8.7 and 2.5 respectively according to Table 13.4). The consistent use of normatives, or planning norms, appears to be behind this stability. An example of the standardized character of retail planning is provided by the 1981-1985 five-year plan which provided four standard dimensions (1500m^2, 2500m^2, 3500m^2, and 7000m^2) for furniture stores, depending on demand factors (Fazekas, p. 44).

While no detailed information is available on the growth in the retail volume handled during this period, it must have paralleled to some extent the growth in overall production, which, as had been mentioned, continued unabated. Retail efficiency must have consequently continued to grow. Thus while the total number of employees in retail grew by 13.4% between 1975 and 1980 (from 395,000 employees to 448,000 according to Table 13.4), output of consumer items such as apparel, textiles and shoes, for example, increased by 10.9%, 7.9% and 6.9% per annum, in real terms, between 1976-1980 (Romania; Documents-Events, Agerpress, Bucharest, Romania, January 1981, p. 39). Increases in efficiency in retailing were apparently being achieved. Indeed the annual plan for 1978 had specified that 82% of the planned increase in retail turnover for that year should come from retail productivity (sales/employee) increases, the rest (18%) from increased retail employment (Fazekas, p. 64). To motivate employees to achieve such productivity increases, increases in retail workers' compensations had, since 1977, been tied to averaged increases in turnover/employee. This conformed to the ideological principle of tying compensation to contribution, mentioned previously, since compensation was thus directly tied to the amount and quality of work performed.

Data on urban/non-urban store size differentiation are presented in Table 13.5. Rural outlets were in all cases considerably smaller, handling undoubtedly reduced varieties and assortments of goods. Interesting is the approximately equal average store size of specialized outlets in urban areas. Much greater variation in average outlet size is apparent in the rural areas, presumably in response to vary-

TABLE 13.5. Retail Distribution (1977) by Category and Type of Product Handled

	Urban			Rural		
Specialized Units	Units	Total Area (1000m²)	Area/Unit* (m²)	Units	Total Area (1000m²)	Area/Unit* (m²)
Processed Food	11593	1243	107.2	5168	333	64.4
Vegetables, fruits	2052	186	90.6	1049	48	45.7
Textiles, shoes	5015	544	108.5	1536	137	89.2
Metal & Chemical products	8687	831	95.7	3931	285	72.5
Books, paper products	961	96	99.9	888	37	41.7
Nonspecialized units						
Department stores	291	323	1110	2272	383	168.6
Variety stores	738	69	93.5	7847	372	47.4
Service units						
Restaurants	1916	714	373	931	190	204
Restaurants - canteens	920	227	247	53	13	245
Druggists	945	122	129	925	54	58.4

Source: Fazekas (1978), p. 42.

*Computed by the author.

ing demand conditions. The extremely large size of newly constructed urban department stores should be noted as well. Planned construction of such hyper-store type outlets envisioned 24 such units by 1980 to range in size from 800m² to 6000m² per unit (Fazekas, p. 44). It was clearly hoped to reap considerable economies of scale from extra large and highly intergrated retail outlets.

The data provided appear to indicate that the distributional principles of planning, equity and efficiency were indeed adhered to in practice.

CONCLUSIONS

As was pointed out, highly centralized planning and administration continue to characterize the Romanian distribution system. The scope available to the authorities for planning the distribution structure is underlined by the fact that fully 52% of the total area of trade establishments in 1978 had been in existence less than 10 years by then (Fazekas, p. 42). Much of such additions in retail facilities involved installations of retail facilities in ground floors of new residential blocks or the renewal or rebuilding of central shopping zones in major urban centers (Ibid, p. 42). It appears that Romania-planners were aware of the need to provide a distribution system that would be in line with the economic and social development goals of the economy while attempting at the same time to stay within the constraints within which such development was permitted to take place.

A number of conclusions can be drawn from the materials presented. On the positive side of the ledger the experience of Romanian-style planning appears to demonstrate that rapid structural changes, involving in this case distribution, may effectively be implemented in this fashion. Centralized planning would appear to permit such development to occur in a balanced fashion as well. Furthermore, planners' prerogatives regarding equity and physical distribution related efficiency can, and apparently have been effectively implemented through the application of central means.

On the negative side we must list such dysfunctional phenomena as overbureaucratization and diseconomies of scale, affecting primarily retail service levels, that may be presumed to have accompanied the process. Incentives and initiative at all distributional levels were severely limited as well, because of the requirement to observe ideological strictures. And, most important, planners' prerogatives were seen to clearly supercede consumer prerogatives. Thus, to the extent that "scientifically determined" consumer needs would differ materially from actual existing consumer needs, satisfying the former would clearly leave the latter largely unsatisfied. Romanian planners were apparently hoping to be able to educate consumers to become rational and "scientifically" minded, in contrast to conditions prevailing in the West, where waste and irrationality were thought to be running rampant, thus reducing likely gaps existing between the two need states. But this could clearly not be achieved in the short run.

Rapid advances in developing a network appropriate to the desires of planners and policy makers appear to have been made in Romania. It remains to be seen whether the authorities will be successful in their quest for a scientific determination of rational consumer needs acceptable to consumers, and for an appropriate structure of distribution of such needs. Their record will clearly affect the attractiveness of the Romanian distributional experience for other change oriented economies, both within the East-bloc and outside it.

REFERENCES

Ceausescu, Nicolae (1978). *Proceedings of the National Conference of Employees in Trade and Tourism*. Bucharest, Romania. Economic Institute for Trade and Tourism, June 27-28.

Ceausescu, Nicolae (1981). *Romania On the Way of Building Up the Multilaterally Developed Socialist Society*. Bucharest, Romania: Meridiane Publishing House.

Der Neu Weg. (1982). (official government German daily), Bucharest, Romania, December 21.

Fazekas, Janos (1978). *Proceedings of the National Conference of Employees in Trade and Tourism*. Bucharest, Romania: Economic Institute for Trade and Tourism, June 27-28.

Goldman, Marshall (1963). *Soviet Marketing, Distribution in a Controlled Economy*. NY: Macmillan.

Law No. 3/1972 Home-Trade Law, Council of State, Editorial Board of the Official Bulletin of the Socialist Republic of Romania, Bucharest, Romania, 1972, 1-35.

Naor, Jacob (1982). *Economic Reform Making in Romania – The 1978 New Economic Mechanism*. Research paper, Equipe de Recherche sur la firme et l'Industrie, Universite de Montpellier, Montpellier, France, 10-23.

Naor, Jacob (1984). "The Conceptual Basis for Public Marketing Policy in Romania – A Socialist Approach." *Proceedings*, 1984 American Marketing Association Summer Marketing Educators' Conference, Chicago, IL, August 12-15.

Samli, A. C. (1978). *Marketing and Distribution Systems in Eastern Europe*. (Praeger Special Studies Book) XXIV, Holt, Rinehart and Winston.

Tsantis, Andreas C. & Pepper, Roy (1979) Coordinating authors, *Romania – The Industrialization of an Agrarian Economy Under Socialist Planning*. The World Bank, Washington, DC, 1-5.

SECTION V: SPECIAL INTERNATIONAL MARKETING TOPICS

Chapter 14

Marketing Ethics and Social Responsibilities of Marketing

Julius O. Onah

INTRODUCTION

Marketing and marketing ethics are not new in Nigeria. What is new is the codification of nationally accepted ethics of conduct for marketers in Nigeria. In the olden days our forefathers had traditions, taboos and values which guided traders and businessmen. But today these traditional values and taboos of our forefathers seem to be in conflict with the values and beliefs of "modern" Nigerians.

Unethical business practices are widespread in Nigeria today. Smuggling, hoarding, profiteering, bribery, corruption and other sharp business practices are common day occurrences in the country. These and many others indicate the absence of business ethics in Nigeria. The word "business," because of these practices, has become a common "umbrella" for any bad business behavior. "Business" is associated with bad business practices. It is seen as a way of cheating others.

Business and marketing profession, in particular, the way it is being practiced today in Nigeria is giving Nigeria a very bad image both locally and internationally. Locally, we can, in our usual way, dismiss the seriousness of unethical marketing practices by the common saying—"that is Business" but can Nigeria as a nation, afford to dismiss same within the international community? The purpose of this chapter is to discuss the various dimensions of marketing ethics and social responsibilities of marketing.

DEFINITIONS

What is ethics? Before we examine some of the unethical behaviors in business in Nigeria it is pertinent that we define the word ethics. Ethics is defined as the philosophy of moral values or moral norms; that is normative ethics. In this case, we can argue that ethics does not belong to empirical science but to philosophy. Ethics could also be regarded as the study and philosophy of human conduct with emphasis on the determination of right and wrong.

Kotler postulated that ethical philosophies can be judged either by the act itself (moral idealism). the actor's motives (intuitionism) or the act's consequences (utilitarianism).[1] Moral idealism restrains the marketing manager from spying on competitors, deceiving customers and other unethical acts. Such managers may refuse to let the end justify the means and thus derive a feeling of correct conduct.

Intuitionism allows individual managers to sense the moral gravity of each circumstance. The manager's sense of justice forms the core of intuition. Utilitarianism deals with the consequences of ethics. If the result of good and bad acts represent a net increase in society's happiness or at least not a net decrease, the act is considered right.

If ethical philosophies, based on concept of the good life of the public and the relation of one's welfare to others, can be worked out, marketers would appreciate the soundness of marketing ethics.

UNETHICAL BEHAVIORS IN BUSINESS

There are many behaviors in business, today, which tend to run counter to our beliefs, values and norms. In a study carried out by Ferrell et al.,[2] to examine the relationship between the ethical behavior of individual managers, peers and top management, it was discovered that behavior is more ethical in some situations than others. The study also revealed that the following behaviors in organization are unethical:

1. passing blame for errors to an innocent co-worker:
2. falsifying time/quality/quantity of reports;
3. claiming credit for someone else's work;
4. padding an expense amount more than 10%;
5. dividing confidential information;
6. pilfering company materials and supplies.

The respondents also indicated other unethical behaviors which they themselves participated in to include:

1. taking extra personal time (lunch hours, breaks, early departures etc.)
2. not reporting others' violations of company policies and rules.
3. doing personal business on company time.
4. using company service for personal use.

Onah and Osuji[3] in their own study identified some trade malpractices in Nigeria such as falsification of scales of measure, deceptive labelling and advertising, mixing products with other foreign bodies in order to increase the volume, and so on.

Other writers have shown that unethical business practices are widespread in Nigeria today. There are smuggling, hoarding and profiteering. Bribery and corruption abound. Discrimination and nepotism exists in the grant of distributorship and agencies. Ifedi went further to add that "seeking preferential treatment through lavish entertainment, kick-backs to purchasing officers, pay-offs to government officials, price-rigging between suppliers and contractors and collusion in contract bidding, underbidding, with substitution of inferior workmanship and materials are rampant with some companies."[4]

These unethical behaviors have debilitating influence, not only in the pursuit of corporate policies but also on the public image of the country. There is, therefore, a great need for marketing ethics.

MARKETING ETHICS

What is marketing ethics? Is there marketing ethics or is there ethics in marketing? What are the sources of marketing ethics? The issue of marketing ethics has been a controversial one. Some people believe that marketing has no ethics because ethical values are relative and subjective. What is ethical to one person may be unethical to another. Others feel that there is marketing ethics.

Marketing ethics, therefore, is defined as the area of marketing study and thought concerned with defining norms for judgments about the moral consequences of marketing actions. In practice, marketing ethics is concerned with standards of adequate behavior in terms of marketing policies and practices within legal and social constraints at a point in time. The concern of the theorist is with what is acceptable behavior in terms of right and wrong in a culture at a point in time so that decision making can be more successful.

Marketing ethics is part of the general ethics in business. Drucker[5] referred to ethics as everyday honesty. He stated that the basic rule of professional ethics (ethics of responsibility) is "primum non nocere," meaning "not knowingly to do harm." Businessmen should not cheat, steal, lie, bribe or take bribes nor should anyone else. Markin[6] defined ethics as the study of right and wrong. Ethics, therefore, embraces a series of perceptions of the rightness involved in daily issues. However, this assertion lacks absoluteness as ethics is merely left to the subjective judgement of individuals. Marketing ethics is concerned with standards for decision-making and right conduct of marketing activities.

Conducts considered unethical to the consuming public include bribery, dishonest communications and marketing of potentially dangerous products. The ambit of marketing ethics covers distribution relations, advertising standards, customer service, pricing and product development.

For those who argue that there is no marketing ethics, is the market a jungle? No, it has rules because of men. The market builds on values—intrinsic and extrinsic. What are the rules? The rules in the market place are simple:

—Reciprocity or quid pro quo
—There must be voluntary mutuality of agreement.
—But the market is not a sufficient moral community
—The market only respects exchangers and not men.

There are, of course, some problems of ethics in marketing, emanating from the norms to be used, i.e., which norms are to be used as standard that would be acceptable to all.

The problems arise in application of norms to specific situations, and in the difficulty of agreeing upon norms or belief decisions in a pluralistic free society.

The Evolution of Marketing Ethics

As we have indicated earlier, there has been a lot of argument about whether or not there is ethics in marketing. Some marketing managers feel they have no grounds on which to base their decisions. But Bartels[7] asserted that the basis of ethical decision for marketing managers evolved through four major stages of sensitivity:

Ethics of self-interest. This is the lowest level of sensitivity to ethical obligation and entails little or no consideration to the expectations of others. Businessmen operate and seek profit for themselves to the exclusion of others' interests. They assume no obligation to competitors, customers, employees, dealers, government or the community. "Laissez Faire," "Caveat emptor" and "the public be damned" are the slogans of ethics of self-interest. Even then there was ethics but the businessman was more concerned with himself than with any other party.

Ethics of compulsion. With the awakening of social conscience, society compels obligation on business participants. Laws are formulated and there is no justification for a lower standard of action. Legislations prohibiting certain types of business behavior are enacted. Companies are thus forced to embrace social responsibility.

Ethics of compliance. As an economy develops, businessmen organize a movement for voluntary formulation of codes of ethics through trade groups. Individual firms state their own ethical tenets. Compliance with acknowledged obligations to various other parties in the socio-economic process are enforced. The obligations represented standards where no laws existed or elevated the legal minimum with a willingness for fuller discharge of responsibilities.

Ethics of conviction. Ethics also evolve from personal conviction. Personal convictions arise from an integrated sense of social and personal values and from respect for law, honesty, fairness and the like. Such ethics bring religious concepts to bear on business relationships and thus interpret men's obligations to men.

Most ethical decisions do face economic limitations. For instance, if there is no profit for a period, the expectations of owners may be denied. If working capital is low, creditors may be required to wait for payment. If costs increase, changes may be made in the product and service normally expected by the customers. Ethical decision is therefore a moral decision impelled by social sanction but modified by economic exigency.

Contents of Marketing Ethics

There is a need for the formulation of marketing ethics in Nigeria. In doing so, marketing ethics could be drawn from the following sources:

1. *Personal Conscience*: We, as Nigerians, have personal conscience. This personal conscience is molded and formed by the ethical traditions of our society. For example, by our tradition, it is unethical to steal or to sell a dangerous drug as food.
2. *The Law and its Corollary*: Our marketing ethics could be drawn from the law of the land. There are today many laws guiding business operations in Nigeria.
3. *Organization Structure and Procedures*: These ensure the interjection of the ethical component into decision making through a system of checks and balances.
4. *The Marketplace:* Our traditional marketing concept is that of

the marketplace which exerts its own ethics on buyer and seller.

5. *Professional Knowledge*: The business and technical expertise which allows one to know what is good for someone else. Professional bodies as sources of ethics do formulate code of conduct which members are bound to follow.

The Nigerian Marketing Association (NIMARK), for example, has a code of practice which members of NIMARK are obliged to follow. Having been worried about the practice of marketing in Nigeria, the Nigerian Marketing Association has come up with a code of practice for its members (see Code of Practice of the NIMARK – Appendix I).

Ethical concept is a two-way communication process. On one hand, we communicate to the consumer through (a) the quality of our products, (b) our packaging and labelling, and (c) our advertising and promotion. On the other hand, the consumer communicates to the seller through (a) repeat purchase of his products, (or the lack thereof), and (b) the myriad of techniques of commercial research – controlled perception tests for appearance, and flavor, label and price tests, retail store tests, and full scale market tests. Any distortion which we consciously introduce in this two-way communication process becomes a violation of marketing ethics.

SOCIAL RESPONSIBILITIES OF MARKETING

Marketing, like medicine, is concerned with people. While medicine is concerned with prolonging life, marketing is concerned with making life worth living by providing satisfaction to people. Marketing, being a behavioral science has some social responsibilities to perform in the society.

Social responsibility of marketing is the obligation of marketing in seeking to accomplish its objectives, to formulate and execute policies as well as make decisions within the constraints of society's values, aspirations and objectives.[8] The constraints may be legal, as expressed in the objectives of the society.

The concept of social responsibility requires businesses to consider their actions within the framework of the whole social system.[9] For the manager, it means realizing that the business system does not exist alone and that a healthy business system cannot exist within a sick society. The marketing executives must conduct their activities in a socially desirable manner in order to justify the privilege of operating in a free economy. It is rather appalling to observe that big organizations which shout the doctrine of social responsibility adeptly practice private greed. Firms make huge profits but impose on society a large part of the costs of the pollution and environmental damage they cause. These adverse effects of business are what Markin[10] referred to as externalities. They are the costs and benefits which accrue to others as a result of a market transaction between two parties. For example, children who do not get any benefits from beer and wine bottles could be injured by these bottles.

Iyanda[11] identifies the social responsibility of marketing in relation to goods and services as:

1. making its advertising claims justifiable in relation to the cognitive attributes of the product;
2. ensuring the accuracy of information transmitted to both producers and consumers;
3. being responsive to post-sale information and service requirement of consumer and devising just means of handling post-purchase cognitive dissonance;
4. operating at a most efficient level in terms of time and cost;
5. ensuring that costs are fair and equitable in relation to services performed.

A MODEL FOR SOCIAL RESPONSIBILITIES OF MARKETING

"Marketing has been widely criticized for its failure to contribute more to the solution of social as well as economic problems." In discussing the social responsibilities of marketing we shall look at marketing responsibilities to the consumer, product, price, place, promotion and society. That is, we shall adopt a (C – 4PS – S) model approach.

Consumer Responsibility

The first responsibility of marketing is satisfaction to the consumer. What the consumer is out in the marketplace to buy is "satisfaction" and not a physical product per se. Therefore, social responsibility of marketing should aim at providing satisfaction to the consumer.

As many Nigerian consumers are uneducated, marketing should, therefore, aim at educating them on the basic consumer rights.[12]

1. The consumer must have protection from fraud, deceit, and misrepresentation.
2. He must have access to adequate information to make an intelligent choice among products and services.
3. He must be able to rely on products working as represented.
4. He must have the right to expect that his health and safety will not be endangered by his purchase.
5. Our marketing system must provide him with a wide range of choice to meet individual tastes and preferences.

At this stage of our economic development, the task of marketing is still that of conversion and stimulation. Consumer Education should be the responsibility of marketing. The age of "Caveat Emptor" is gone and services given out to them. By educating the consumer, marketing shall be increasing the standard of living of the people.

While the Nigerian Enterprises Promotion Act (1970) made provision for indigenization of enterprises in Nigeria, not much effort was made toward consumer protection. We propose that the Committee on Ethical Revolution should look into this. While Nigerians are taking over alien businesses under schedules I and II adequate provisions should be made for consumer protection. There should be:

- a consumer protection law with provision for consumer class action.
- a consumer product testing law
- expanded Government consumer education activities

– a new look at guarantees
– an office of consumer affairs in the office of the president.

These proposals will go a long way in making the industries sit up to their responsibilities.

Product Responsibility

One of the social responsibilities of marketing is to produce products and services for all the segments of the society. In doing so marketing should ensure that the product is safe for consumption. It is now believed by many social scientists and policy-makers that for new product development to be successful, social and environmental factors must be considered.[13]

The quality of products and services is a social responsibility of marketing. Onah and Osuji discovered in their study that some businessmen and women add foreign objects like stone, chalk, and carpenter's dust to swell the products they are selling. These objects do affect the quality of products. More recently, some of our beer companies are being accused of adding more water to their beer after the Excise duty officials have completed their inspection. We, therefore, propose that the Standard Organization of Nigeria (SON) should certify all new products being produced in Nigeria because quality is the key to consumer satisfaction.

Pricing Responsibility

The interference of Governments of Nigeria in pricing is a clear indication of lack of social responsibility of marketing towards pricing. The price control and resale price maintenance introduced by the Federal Government of Nigeria and clear examples.

It is the social responsibility of marketing to charge what the "traffic can bear." Optimal pricing strategy is what is recommended for Nigeria. That is about 10-20% mark-up. This strategy would enable industries to earn profit that is acceptable to both industries and consumers.

The pricing system needs to be reviewed. The system of higgling and haggling is still the practice and even where the Government has recommended some optimal price, businessmen and women waive the policy by allowing the customer to haggle. It is our recommendation that all retail stores should use "price labels" on their goods and services. A law should be promulgated to enforce the use of price labels.

Place Responsibility

The social responsibility of marketing also includes providing goods and services at the place where they are required. Convenience, therefore, is the prime social responsibility of marketing. Even where an organization uses distributors or agents as is the case in Nigeria, consumer satisfaction through convenience is still the responsibility of the manufacturer.

In a study conducted in 1978, it was found that product availability is today one of the determining factors for purchase decision.[14] Marketing should, therefore, consider product availability as its social responsibility to consumers.

Promotion Responsibility

Two basic consumer rights—"Right to information," and "Right to be heard" are the direct responsibilities of promotion. Our markets are full of products with false claims, deceptive advertising and labeling. For example, some products which are made in Nigeria have on their labels—"Made in Hong Kong," or "Made in Great Britain," given the false message of being foreign-made. There should be a law forbidding such actions. Product packages should be made to carry on them the constituents of the product and the country of origin. Marketing has the social responsibility of telling the truth at all times.

Societal Responsibility

Marketing begins and ends with the consumer. The issue of societal marketing has been a very popular one, though no one has come up with any acceptable solution. The social responsibility of marketing is a very important one.

Societal responsibility of marketing could be summarized to include:

– satisfaction of human needs,
– expansion of social fields, and
– consideration of social impact.[15]

Business must always consider the overall affect of its actions on the society.

BUSINESS ACTIONS TOWARD SOCIALLY RESPONSIBLE MARKETING

Companies whose business practices tend to destroy competition, raise barriers to entry and gain the protection and favors of legislators are not competing fairly. With all the criticisms of malpractices being heaped on business, it is not enough to do some good in the social arena. Business executives should let the public know what they are doing to improve the quality of life as well as the standard of living, instead of hiding their light under a bushel.

Kotler[16] sees five principles businesses can adopt in an effort to achieve socially responsible marketing:

Customer-oriented marketing. The focus of the attention of business should be to serve and satisfy customer needs at a profit. The marketing plans of companies should be designed from a customer point of view.

Innovative marketing. Companies should intensify actions in searching of product and marketing improvements. Any company that overlooks new and better ways of conducting its activities may be competed out.

Value marketing. Marketers that engage in one shot sales promotions, minor packaging changes, or advertising puffery may succeed in raising sales in the short-run. In the long-run, such marketers would discover that less value is added to the consumer and that real efforts are needed to improve the product's features, convenience, availability, information, etc. Companies should, therefore, direct their efforts toward value building marketing investments.

Sense-of-mission-marketing. Company policies must be set in broad social terms; clear definition of objectives gives a good guide toward the performance of business operations.

Societal marketing. Any company striving for socially responsible marketing must make decisions that take into consideration the customers' wants, the company requirements, the consumers' long-run interests and the society's long-run interests. It should be noted that any disregard for the long-run interests of the consumer and the society by business is a disservice to mankind.

SOCIAL RESPONSIBILITIES AND BUSINESS ETHICS

The conduct of business activities by today's businessmen seems to be highly ethical though some exceptions, particularly in Nigeria, exist. There are incidences of false or misleading advertisements, overpriced shoddy goods; bribes to win lucrative contracts; polluting industries and products that harm our health. Our businessmen still engage in unhealthy price fixing schemes with competitors, spread injurious rumors about other companies and introduce built-in self-destructive mechanisms in their products (planned obsolescence), so that their new products can be purchased. Companies indulge in deceptive packaging of products, elude government regulations and carry out industrial espionage on competitors.

Since a responsible marketing action is the one that does not arouse the executive's "sense of injustice,"[17] if the aforementioned cases arouse sense of injustice, they are unethical and irresponsible actions. Carr[18] argued that ethics of business are game ethics, different from the ethics of religion. According to him, it is a game that demands both special strategy and an understanding of special ethics. Following Carr's school of thought, it seems that violations of ethical ideals of society which are common in business are not necessarily violations of business principles. Perhaps, this view is in line with Robbin's[19] argument that "espionage in business is not an ethical problem, it is an established technique of business competition." He went further to say that "so long as a businessman complies with the laws of the land and avoids telling malicious lies, he is ethical."

These views, therefore, make controversial the interpretation of what is right or wrong in business vis-à-vis the ethical ideals of the society. Reconciling personal integrity and high standards of honesty required of a marketing manager with the practical requirements of business implies that ethical standards be assessed subjectively. Private ethical standards often run counter to business ethics. For example, a manager can be ordered to deny promotion to a man who deserves it; to give an employee of long standing a quit notice; to prepare advertising that he believes to be misleading; to conceal facts that he feels customers are entitled to know; to cheapen the quality of materials used in the manufacture of an established product; to sell as new a product that he knows to be rebuilt; to exaggerate the curative powers of a medicinal concoction or to coerce dealers. These ethical dilemmas are enough to subject a manager to contemplation. Since contemplation bakes no loaves, he either dances to the rule of game ethics (business ethics) or loses his seat in the company.

Austin[20] pointed out that the true responsibility of business management is to make some appraisal of the social effects flowing from its strategic policy decisions and technological advances. There are ethically difficult situations where management is aware of the adverse consequences of the technological advances of its company and yet remains helpless, for example, a manager of a cigarette company who knows the link between cigarette smoking and cancer but cannot stop manufacturing cigarettes.

CONCLUSION

Ethical issues and social responsibilities of marketing require urgent attention now more than ever before mostly because of the need for consumers to be protected and the society, in a wider context, to benefit from an improved business climate. The producers and consumers alike should cooperate to ensure that the interests of each group are adequately protected with or without government intervention.

APPENDIX 14.1

NIMARK Guides to Professional Conduct

Professional conduct involves the marketing man's own sense of integrity and his professional relationship with those to whom he renders services, with his employer, with other members of the profession, the community, the nation and the world at large. In all these relationships every member of the profession is concerned with his own behavior and, because the good name of the profession is the concern of all its members, with the behavior of his colleagues as well.

Good marketing practice seeks to make fair and honest profit by meeting the needs of the customer. It is in the interest of every member and of the profession that marketing malpractices of any kind be avoided or checked. While the law has its role in this regard, the best form of protection for members of the profession and the consumers of their services lies in self-restraint. It is only through such self-restraint or internal "policing" that the long term interest of the profession and its members can be maintained.

CODE OF PRACTICE

Marketing's Professional Responsibility

The professional marketing man or woman has responsibilities to his employers, to customers—both ultimate and intermediate—to his colleagues, to the profession, to the public and to society at large. The N.M.A. requires its members, as a condition of membership to recognize these responsibilities in the conduct of their business and to adhere to the Code of Conduct. All members shall be answerable to the (Governing Body) of the N.M.A. for any conduct which in the opinion of the (Body) is in breach of this Code and the (Body) may take disciplinary action against any such offending member.

Professional Duty

(a) A member shall at all times act in a manner to uphold dignity of the marketing profession and to fulfill its responsibility to the public.

(b) The member will bear in mind that the marketing man acts as an expert when he gives marketing advice, and he will give such advice only when he is qualified to do so. He shall need to ensure that all who work with him or for him have the appropriate levels of competence for the effective discharge of the marketing tasks entrusted to them and where any shortcomings might exist, he will seek to ensure that they are made good as speedily as possible.

(c) The member shall not provide marketing service for or associate professionally with any person or organization where there is an evident possibility that his service may be used in a manner that is contrary to the public interest or the interest of his profession or in a manner to evade the law.

Relationship to Client or Employer

(a) The member shall act for each client or employer with scrupulous attention to the trust and confidence that the relationship implies and will have due regard for the confidential nature of his work.

(b) The member will exercise his best judgment to ensure that the recommendations made by him or under his direction are based on sufficient and reliable data, that any assumptions made are adequate and appropriate, and that the methods employed are consistent with the sound principles established by precedents or common usage within the profession.

Honesty

A member shall at all times act honestly and in such manner that customers—ultimate and intermediate—are not caused to be misled. Nor shall he in the course of his professional duties knowingly or recklessly disseminate false or misleading information. It is also his responsibility to ensure that his subordinates conform with these requirements.

Relations with Other Members

The member shall conduct his professional activities on a high plane. He shall avoid unjustifiable or improper criticism of others and will not attempt to injure maliciously, the professional reputation of any other marketing professional. He shall recognize that there is substantial room for honest differences of opinion on many matters.

Conflict of Interest

(a) In any situation in which there is or may be a conflict of interest involving the member's marketing service, whether one or more clients or employers are involved, the member shall not perform such marketing service if the conflict makes or is likely to make it difficult for him to act independently.

(b) A member shall use his utmost endeavor to ensure that the provisions of his Code and the interest of his customers are adequately and fairly reported to his company in any circumstances where a conflict of interest may arise.

(c) A member holding an influential personal interest in any business which is in competition with his own employer, shall disclose that interest to his employer.

(d) A member having an influential personal interest in the purchase or sale of goods or services between his own company and another organization shall give company prior information as to that interest.

Confidentiality of Information

(a) A member shall not disclose, or permit to be disclosed to any other person, firm or company, any confidential information concerning a clients' business without the written consent of the customer except where required by law.

Solicitation of Business

The member shall neither engage in nor condone any advertising of any other activity which can reasonably be regarded as being likely to attract professional work unfairly, or where the one, form, and content are not strictly professional.

Relationships with Other Professional Groups

The member shall in all his associations with any other professional associations or groups uphold at all times the image and integrity of the marketing profession.

Source: Marketing News, A Publication of the Nigerian Marketing Association.

REFERENCES

1. Philip Kotler, *Principles of Marketing*, Prentice Hall Incorp., Englewood Cliffs, New Jersey, 1980, p. 669.
2. O.C. Ferrell and K. Mark Weaver, "Ethical Beliefs of Marketing Managers," *Journal of Marketing* Vol. 42. No. 3, July 1978, p. 71.
3. J. O. Onah & L. O. Osuji, "Trade Malpractice in Nigeria," *Journal of Business Management*, Vol. 3, No. 1, Jan/Feb., 1982, p. 44-46.
4. CHUMA Ifedi, "Business Ethics in Nigeria," *Daily Times*, Wednesday, March 23, 1983, p. 7.
5. Peter Drucker, *Management, Tasks, Responsibilities, Practices*, William Heinenmann Limited, London, 1974, p. 366.
6. Rom. J. Markin, *Marketing*, John Wiley and Sons, New York, 1979, p. 672.
7. Robert Bartels, "A Model for Ethics in Marketing," *Journal of Marketing*, Vol. 31, No. 1, Jan. 1967, p. 22.
8. Olukunle Iyanda, "The Social Responsibility in Marketing," *Management in Nigeria*, Aug. 1978, p. 49.
9. W.J. Stantan, *Fundamentals of Marketing*, McGraw-Hill, Tokyo, 5th Ed., 1978, p. 49.
10. Rom. J. Markin, *Op. Cit*, p.673.
11. Iyanda, *Op. Cit.*, p. 56.
12. Robert J. Lavidge, "The Growing Responsibilities of Marketing," *Journal of Marketing*, Jan. 1970, Vol. 34, No. 1, p. 25.
13. Mauric H. Stans, "Marketing and Consumer Interest," *The Marketing News*, January 1, 1970, p. 3.

14. Dale L. Varble, "Social and Environmental Considerations in New Product Development," *Journal of Marketing*, October 1972.

15. Nkeora & Associates, Marketing Planning and Strategy. (unpublished report).

16. Andrew Lakas, "Societal Marketing: A Businessman's Perspective," *Journal of Marketing*, Vol. 38, No. 4, October 1974, p. 2.

17. Philip Kotler, p. 666.

18. James M. Patterson, "What Are the Social and Ethical Responsibilities of Marketing Executives?", *Journal of Marketing*, Vol. 30, No. 3 July 1966, p. 14.

19. Albert Z. Carr, "Is Business Bluffing Ethical?" *Harvard Business Review*, No. 1, 102, 1970, p. 102.

20. See A.Z. Carr, *Op. Cit.*, p.104.

21. Robert W. Austin, "Responsibility for Social Change," *Harvard Business Review*, July-August 1965, p. 147.

Chapter 15

Perception of Husband and Wife in Family Decision Making in an Oriental Culture: A Multidimensional Scaling Approach

Oliver H. M. Yau
Leo Y. M. Sin

INTRODUCTION

There has been considerable research in the field of marketing concerning the family decision-making process. Most research is conducted as though individuals make decisions independently of other household members (Kasulis and Hughes, 1984). However, it was found that husbands and wives performed different roles in purchasing decisions. Employing a convenient sample of 50 couples of married college students, Kenkel (1961) discovered that husbands assume an instrumental role in helping their spouse make final purchasing decisions while wives assume an expressive role to decide on color, style, and design which conform with social norms. Davis (1970) further improved the idea of the roles of husband and wife in purchasing decisions. The results of his study rejected the belief that family purchasing decisions took a unidimensional or bidimensional role structure.

He concluded that an "overall" power (dominance) score for the family is not appropriate as consensus between husband and wife can be reached in purchasing products in specific categories. Along the same lines, Scanzoni (1977) also identified changing sex roles in family decision making.

Ferber and Lee (1974) successfully confirmed that among newly married couples there was a high tendency for joint decisions and that wives in American society tended to take the role of family financial officer. Szybillo, Sosanie and Tenenbein (1979) also supported this finding by indicating that husbands and wives did exert, to a great extent, independent influences on the household purchase decisions, while Cosenza and Davis (1980) reported that the locus of familial control appears to shift when the wife becomes employed. However, different findings were found. Sheth (1984) explained that greater incidence of working wives has created time pressure within the family, encouraging individual decision making for many products that might ordinarily be purchased on a joint basis. Filiatrault and Ritchie (1980) revealed that husbands tend to dominate decision making more in family decision-making units (husband, wife, and children) than in those where no children are present.

While research efforts were made on joint decisions, the focus of study has been shifted to an international comparison of family decisions. Green, Verhage and Cunningham (1979) were amongst the earliest to compare how American and Dutch consumers differ in household purchase decisions. It was found that substantial and significant differences exist between the two countries. Wives in the U.S. play a more autonomous role than the Dutch wives in family decision making but purchase decisions in Dutch families tend to be made on more of a joint basis than in U.S. families. Along the same lines, Tan, Teoh and McCullough (1986) compared responses of dyads of husbands and wives on buying decisions in a cross-cultural analysis between Singapore and the United States. Discrepancies between the two were found.

However, perceptual differences between husband and wife were related to the family's values orientation in Singapore but not in the U.S. These research findings seemingly suggest the existence of cultural difference and support the need for contextualization of consumer behavior in a specific culture as advocated by Engel (1985). Therefore, this chapter tries to achieve the following objectives in reference to an oriental culture:

- investigate husband-wife influence at different stages of the family purchasing decision process;

– measure the congruence of role perception between husbands and wives at different stages of the family purchasing process; and
– investigate if differences in the perception between family purchasing decisions exist between working wives and non-working wives.

METHODOLOGY

The Sample

In order to minimize the costs and to increase accessibility of respondents, sampling of the research consisted of several steps. In the first step, sample size was determined to be 75 couples of husbands and wives. In the literature on this type of research, this size was considered to be adequate (Davis and Prigaux, 1974). In the second step, the multi-stage cluster sampling method was employed. First, eight geographic regions were randomly drawn. Then with the help of the "list of buildings in Hong Kong" (Wah Kiu Yat Pao, 1985), two blocks of buildings were randomly selected from each region. Similarly, five living quarters were chosen as the tertiary sampling units. If a living quarter had more than one household, then the couple of the households was selected randomly as the unit of inquiry.

Data Collection Method

The personal interview method was adopted to obtain systematic information from spouses in each sampling unit. In order to increase the response rate, a priori notification letter was sent to each household in the sample, informing them of the visit of interviewers. In order to eliminate environmental effect and to reduce the influence from the spouse, two interviewers were responsible to visit a household and interview the couples separately and simultaneously.

Questionnaire Design

The questionnaire was the same for both the husbands and wives. It consisted basically of two parts. The first part was a two-way table in which twenty household purchasing decisions (see Table 15.1) were listed as row elements while the following three stages of the decision process were treated as column elements: (1) problem recognition stage, (2) information search stage (internal and external), and (3) final decision stage.

For each cell, respondents were requested to indicate the major influence (husband, wife or joint decision) in different stages, of the decision-making process. The relative influence of each spouse was determined according to the assumptions of Kenkel (1961). The second part of the questionnaire consisted of seven socio-demographic variables.

The questionnaire was written in Chinese and all interviews were conducted in Cantonese, which is the most commonly spoken dialect in Hong Kong. One aspect of the methodology used in the study deserves comment. Similar to many other studies of family roles in decision making, we have made use of direct questions about the influence of each spouse (husband or wife) at different stages of the decision process, instead of the diary method which keeps track of the influence of each spouse (Davis and Rigaux, 1974). In Hong Kong, the diary method is usually not used partly because the pace of life in Hong Kong is so fast that respondents are really in trouble if they want to keep an accurate record of their influence on purchase decisions, and partly because it is much more expensive than personal interviews.

TABLE 15.1. Twenty Family Purchase Decisions (and Key to Figures 15.2a through 15.5c)

1	Kitchenware	11	Refrigerators etc.
2	Children's Clothes	12	Toothpaste & Soaps
3	Living Room Furniture	13	Outdoors Entertainments
5	Cosmetics and Toiletries	15	Non-prescription Drugs
6	TV, Hi-Fi and Tape Recorder	16	Husband's Clothes
7	Non-alcoholic Beverages	17	Household Cleaning Products
8	Housing Upkeep Items	18	Children's School & Study Program
9	Children's Toys	19	Family Vacation
10	Wife's Clothes	20	First-aid Items

DATA ANALYSIS

Several methods were employed to analyze the data. In order to achieve the first two objectives of the study, multidimensional scaling was used to obtain role perceptions of the twenty household purchase decisions at different stages of the family decision process. Figure 15.1 shows the procedures by which mappings were derived. Firstly, for each stage of the purchase process and for husband and wife respondents, a similarity matrix was computed by the subjective clustering program written by Carmone (SCJC Program, 1975). This matrix shows the frequency of consensus of husband and wife in their role perception. Secondly, this matrix was then input to the MDSCAL-5M program (Smith, 1985) to obtain a configuration of the twenty household decisions. Thirdly, the same matrix was used to group decisions (products) into three clusters by the diameter method of the HICLUST program (Smith, 1985). These three clusters of decisions were labeled: husband dominance decisions, wife dominance decisions and joint decisions. Fourthly, configurations of husbands and wives for each stage of the purchase process were compared by the CFMH program, which provides an index to measure goodness of fit between the two configurations.

RESULTS

The Problem Recognition Stage

Figures 15.2a to 15.4b show perceptions of husbands and wives in the aggregate for the three decision stages. At the problem recognition stage, both husbands and wives have more or less the same classification of purchase decisions. For example, purchase decisions on alcoholic beverages and husband's clothes were grouped as husband dominance by both husbands and wives, while toothpaste and soaps, child-related products, food, and drugs were perceived as wife-oriented decisions. However, when comparing perceptual mappings shown in Figure 15.2a and 15.2b, it is worth noting that husbands have a clearer classification of decisions than wives. The selection of alcoholic beverages was perceived more as a husband-oriented decision at the stage of problem recognition by husbands than wives. The selection of husband's clothes was also husband-

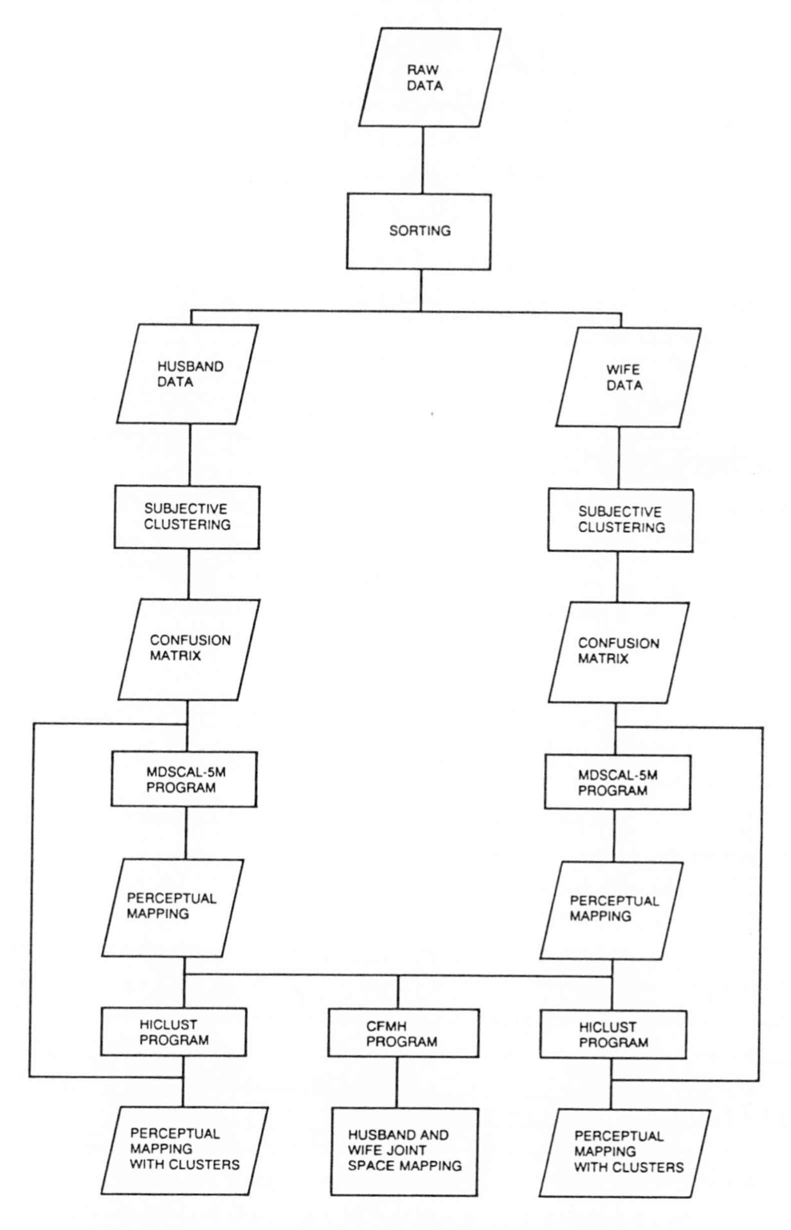

FIGURE 15.1 Procedure for Obtaining Perceptual Mappings

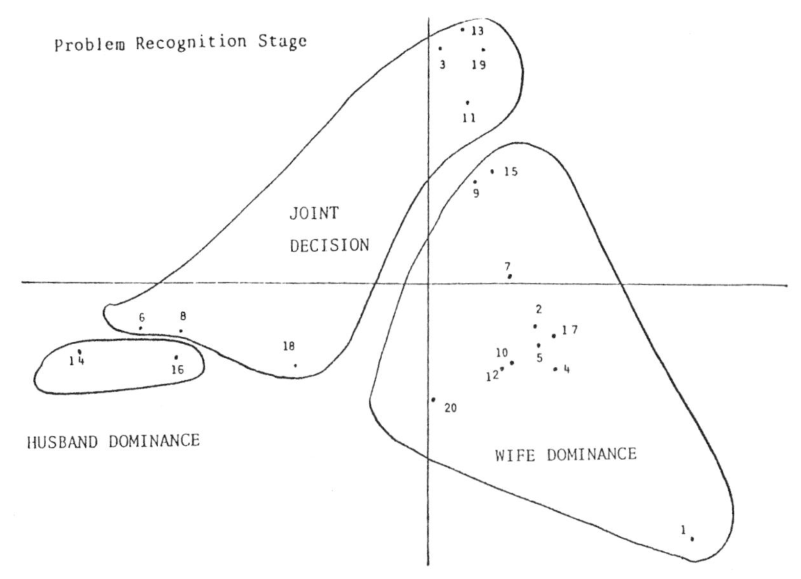

FIGURE 15.2a. Wife's Perception of Family Purchase Decision at the Problem Recognition Stage

dominant. However, the selection of electrical appliances and house upkeep items was regarded as a joint decision by wives and not by husbands.

Figure 15.2c reports the congruence of the configurations shown in Figure 15.2a and Figure 15.2b. The goodness of fit was found to be 0.98 which indicates that the two configurations are very similar. In other words, husband's and wife's perceptions of purchase decisions at the problem recognition stage are very close. Figure 2c also shows that three product decisions are perceived differently at this stage. They are (1) refrigerators (2) outdoor entertainment and (3) TV, Hi-Fi and tape recorders.

The Information Search Stage

With respect to the information search stage (see Figures 15.3a and 15.3b), husbands and wives in the sample have more or less the same classification of purchase decisions. The purchase of children's toys was regarded as a joint decision by wives but as a wife-

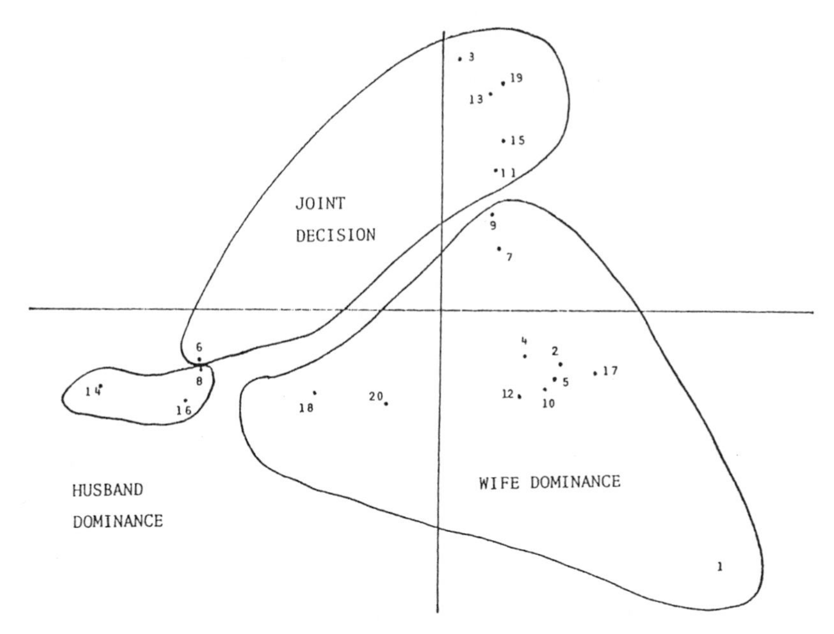

FIGURE 15.2b. Husband's Perception of Family Purchase Decision at the Problem Recognition Stage

dominant decision by husbands. Further, the buying decision of refrigerators was classified as a joint decision by husbands but a wife-dominant one by wives. Hence, it seems that husbands have a more logical and clearer classification of purchase decisions than wives.

Figure 15.3c shows the congruence between husbands' and wives' perceptions of purchase decisions at the information search stage. The goodness of fit index was found to be 0.987 which indicates that there is a small variation between husbands' and wives' perceptions. Two notable differences are the purchase of refrigerators and living room furniture. However, these differences are still within tolerable limits.

The Final Decision Stage

With respect to the final decision stage (see Figures 15.4a and 15.4b), the aggregate perceptions of wives and husbands differ con-

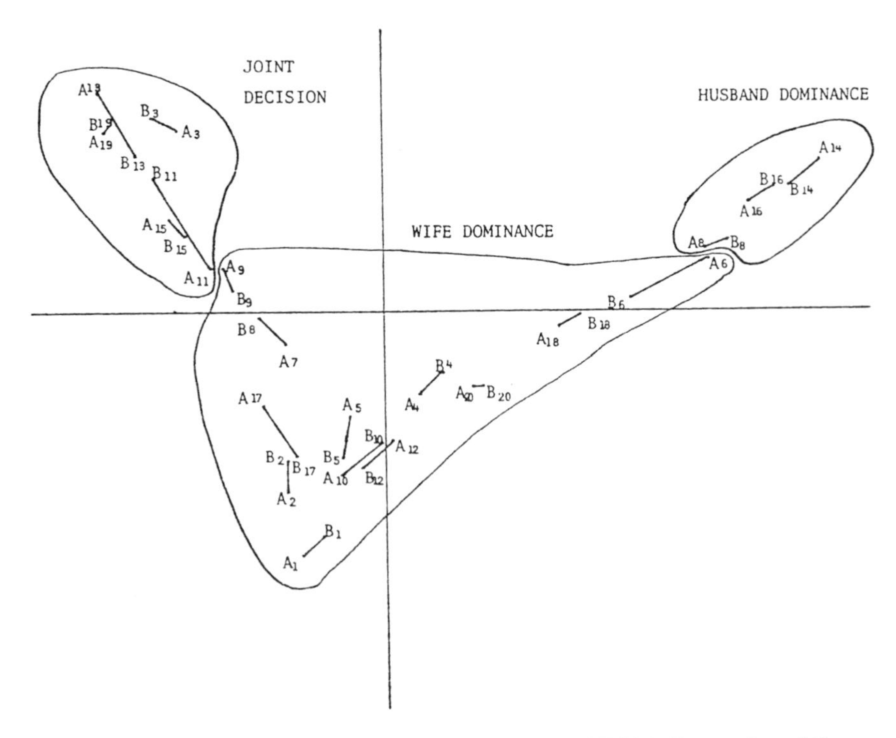

FIGURE 15.2c. Comparison Between Husband's and Wife's Perception of Family Purchase Decision at the Problem Recognition Stage

siderably from those of previous stages. Husbands perceived their influence to be greater and the degree of joint decision making to be less. This finding is not consistent with that of Kasulis and Hughes (1984). The responses of husbands lead to the classification of the selection of furniture, vacation and children's study programs as husband dominant whereas the responses of wives lead to the classification of the selection of house upkeep items as husband-dominant. However, at this stage, it seems that wives have a clearer and more consistent role perception of decisions than husbands, when comparing Figure 15.3a with Figure 15.3b. There are two possible explanations. Firstly, it might imply that husbands in Chinese society still perceive themselves as the head of the family and the source of authority, so that they tended to have greater involvement in

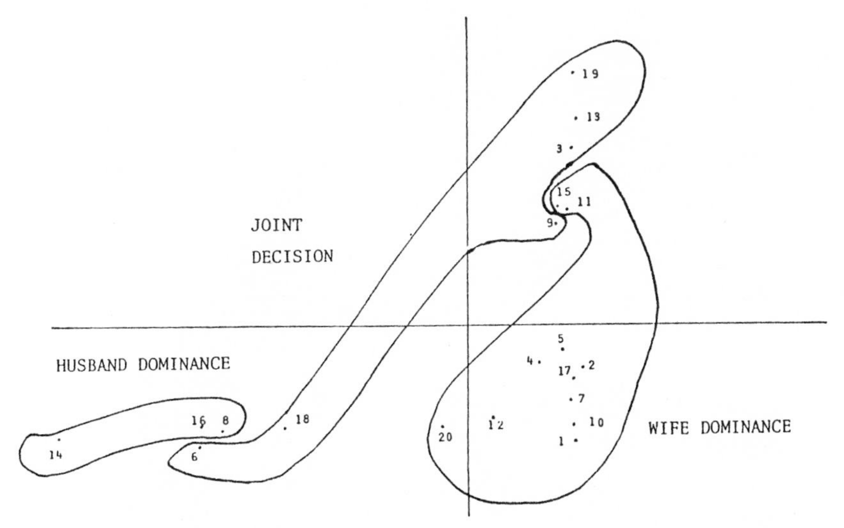

FIGURE 15.3a. Wife's Perception of Family Purchase Decision at the Information Search Stage

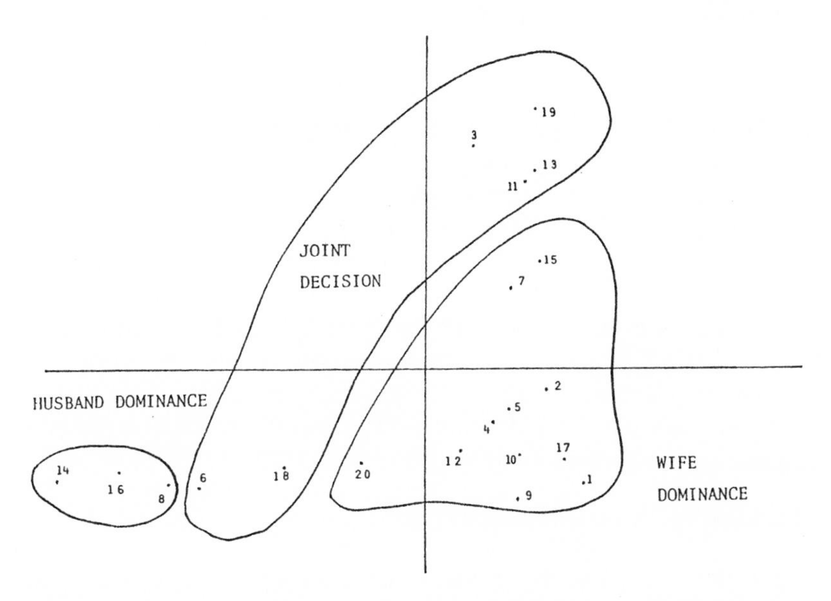

FIGURE 15.3b. Husband's Perception of Family Purchase Decision at the Information Search Stage

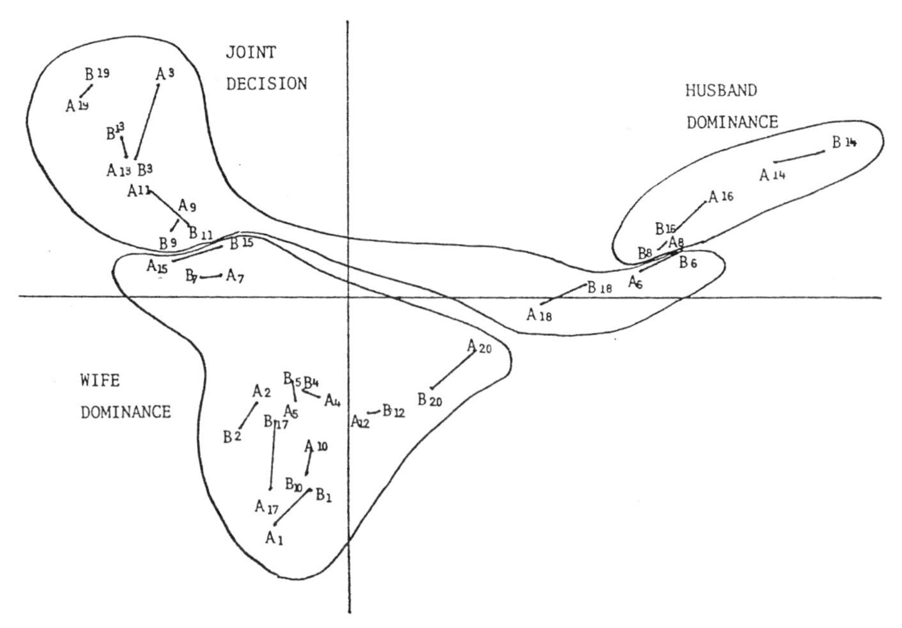

FIGURE 15.3c. Comparison Between Husband's and Wife's Perception of Family Purchase Decision at the Information Search Stage

family purchase decisions. Secondly, this finding could exist because husbands were willing to admit dominance whereas wives were not. If the latter were true, then it is consistent with the finding of Minsinger, Weber and Hansen (1975).

Figure 15.4c shows a joint-space mapping of both husband's and wife's perceptions of the purchase decisions at the final decision stage. With the goodness of fit index being 0.978, it can be concluded that husbands and wives have more or less the same perception of purchase decisions. Some more notable differences in perception are purchase decisions related to (1) non-alcoholic beverages (2) non-prescription drugs (3) food and (4) first-aid items.

Working vs. Non-working Wives

In order to explore the influence of working wives, it is interesting to investigate the differences of perceptions between working wives and non-working wives. It is found that basically the differ-

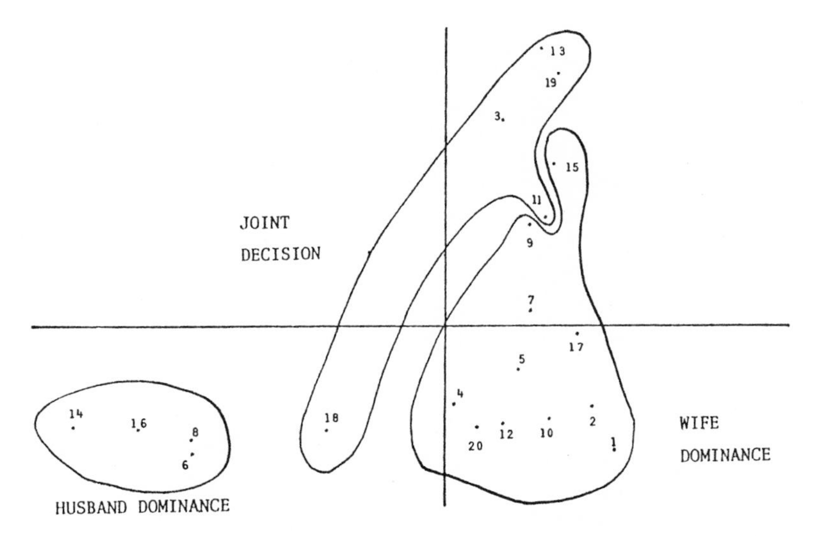

FIGURE 15.4a. Wife's Perception of Family Purchase Decision at the Final Decision Stage

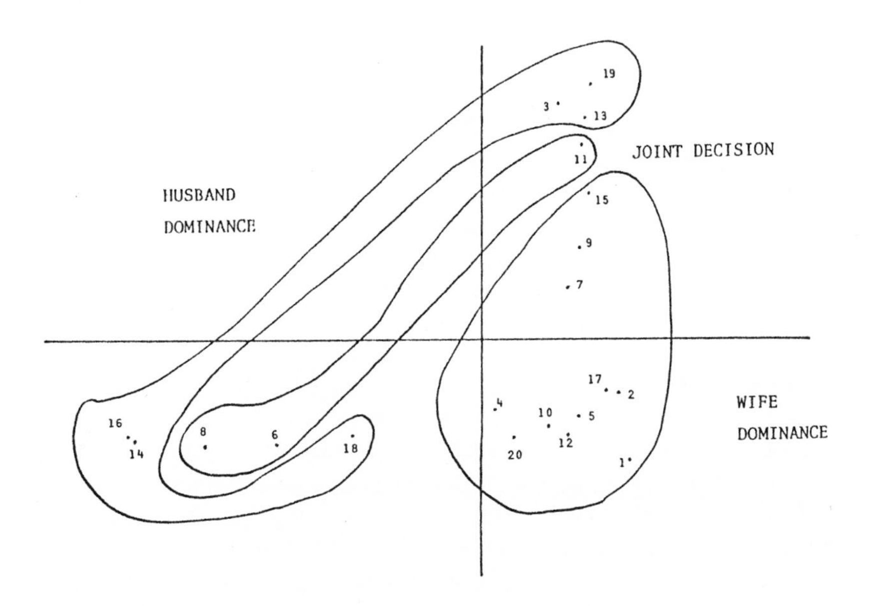

FIGURE 15.4b. Husband's Perception of Family Purchase Decision at the Final Decision Stage

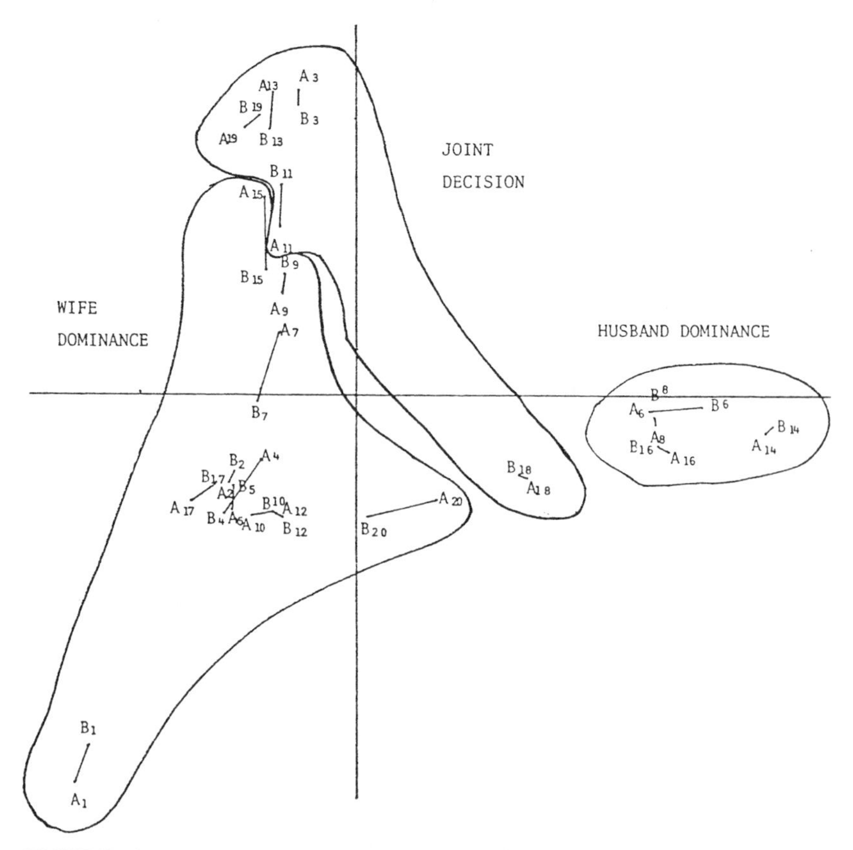

FIGURE 15.4c. Comparison Between Husband's and Wife's Perception of Family Purchase Decision at the Final Decision Stage

ences are very small at the first two stages (not shown in this chapter). One difference worth noting between these two groups at the final decision stage is the selection of kitchenware which was classified as a joint decision by working wives whereas classified as a wife-dominant decision by non-working wives (see Figures 15.5a and 15.5b). Another difference is the selection of food, which is regarded more as a joint decision by working wives than non-working wives. Besides the differences in classification of purchase decisions, the differences of perceptions are obviously large as well. Figure 15.5c shows the joint-space mapping of both husband's and

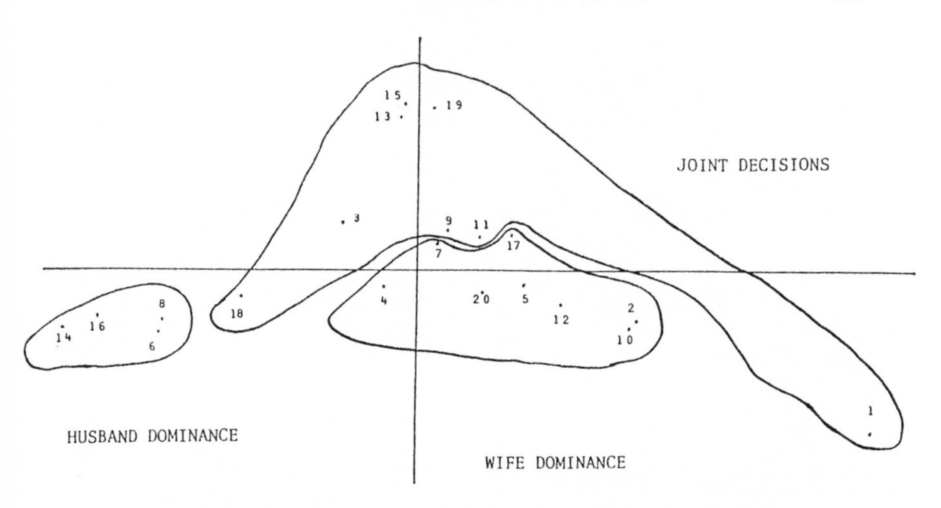

FIGURE 15.5a. Working Wife's Perception of Family Decisions

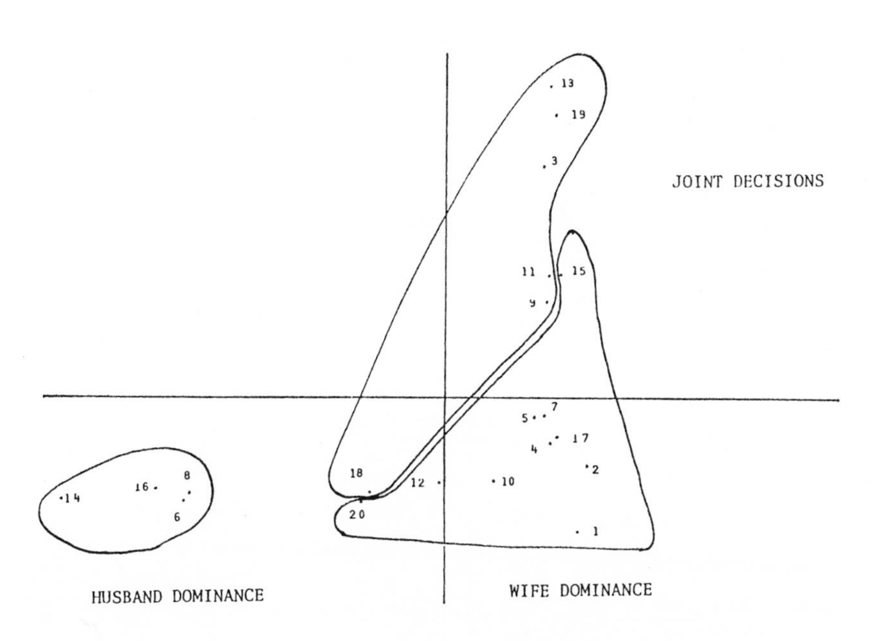

FIGURE 15.5b. Non-working Wife's Perception of Family Decisions

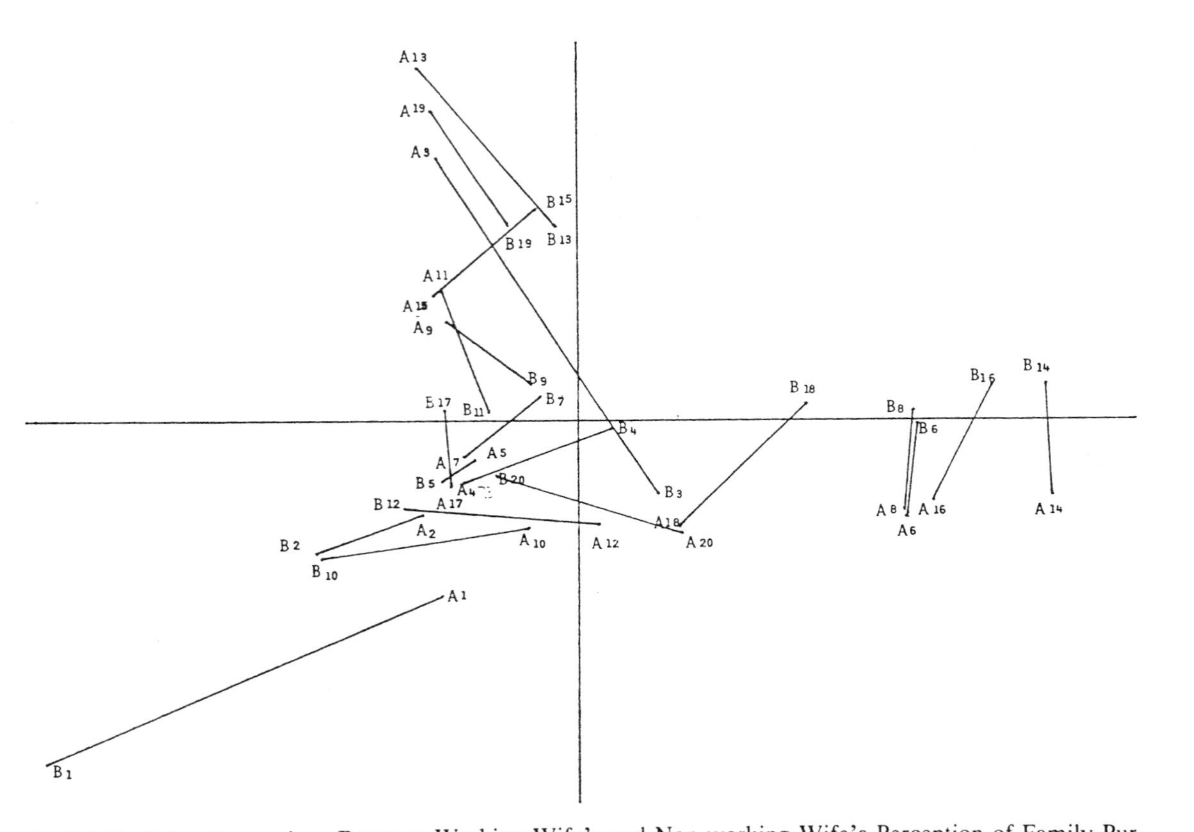

FIGURE 15.5c. Comparison Between Working Wife's and Non-working Wife's Perception of Family Purchase Decision at the Final Decision Stage

wife's perceptions of the twenty purchase decisions. The goodness of fit index was found to be 0.67, which indicates the two perceptual mappings (Figure 15.5a and Figure 15.5b) are not similar. The differences of perceptions are probably due to the working status of wives who become so busy that they perceive cooking affairs as joint decisions.

DISCUSSION AND CONCLUSIONS

A major portion of past research on the relative influence of husbands and wives in various decisions dealt with samples in the States. This research is among the earliest to explore whether similar patterns of husband and wife roles in family purchase decisions can be found in the Chinese culture. It is expected that role differentiation may have different implications for product planning, choice of distribution channels, advertising contents and media. The findings in this research study suggest that a knowledge of decision influence can be helpful in designing marketing strategy for various products. For example, the results suggest that in marketing men's wear, alcoholic beverages, electrical appliances and products related to house upkeep, it seems more effective to direct their appeals to the husband, since he is usually perceived as having the most influence in the choice of these products, especially in the final stage of purchase decision making. On the other hand, the decision-making pattern for drugs, toothpaste, soaps and children's clothes suggests that a more female oriented strategy might be adopted.

In contrast, working wives in the aggregate tend to have a different perception of family purchase decisions from non-working wives (and husbands as well), especially at the final stage of decision making. Findings show that there is a high tendency for some wife-oriented decisions, such as food, kitchenware and first-aid items, to become joint decisions, probably because working wives have less time to make such types of decisions, whereas purchase decisions such as cosmetics, children's clothes and wife's clothes become more wife-dominant. To marketing practitioners, therefore, it is worth noting that potential target markets should first be identified before appeals are made to the appropriate group.

In this research study, no attempt was made to investigate the effect of socio-economic and other demographic variables on attitude and behavior concerning role preferences, and expectations and differences in influence structure in decision making in the Chinese context. Further research is also necessary to determine if such influence might be significantly associated with variation in Chinese culture values with respect to the selection of family purchase decisions.

Chapter 16

A Strategic Stakeholder Approach to the Marketing of Tourism and Recreation Facilities by Local Authorities

Marius Leibold

INTRODUCTION

Marketing of tourism and recreation facilities has become of significant importance for public authority decision-makers in the past few years. With socio-economic and other environmental changes leading to major increases in population discretionary expenditure and leisure time available, the demand for tourism and recreation facilities has increased considerably. This has caused a number of far-sighted public institutions to embark on the planning and development of various facilities, with the view towards achieving an optimum fit between the satisfaction of tourism and recreation needs, and balanced local infrastructure development and community benefits.

The task is by no means a simple one, as tourism (and also, to a certain extent, recreation, especially in a local community context) is a relatively difficult phenomenon to describe. The idea of a "tourism industry" is attractive, but the link between tourism, travel, recreation and leisure is nebulous. However, the phenomenon and its impact, although difficult to measure scientifically, is making itself clearly felt.

In the analysis of, and dealing with, these issues, it seems that public authorities—be it on national, regional or local levels—have only recently realized the importance of scientific marketing. The reason for this may be due partly to incorrect perceptions of the nature of marketing and its constituent parts, and partly to the fact that the issues at stake demand that a new, industry-specific approach to marketing should be adopted. What is clear is that heightened competition between tourism destinations and alternative recreation facilities necessitate appropriate marketing thinking with a sound strategic orientation.

It is against this background that a research project was launched during 1987 with the objective of determining the existing approaches to marketing of tourism and recreation facilities by local authorities, and, if the major hypothesis that such approaches are inadequate proved to be valid, to develop and propose a new and/or more appropriate approach to be adopted. This article briefly outlines the nature and task of contemporary marketing, states the methodology used in the research, and then presents the major findings of the research.*

THE NATURE AND TASK OF CONTEMPORARY MARKETING

Marketing is defined by the American Marketing Association (1985, p. 2) as "the process of conceptualization, pricing, communication and distribution of goods, services or ideas by way of mutually beneficial exchange processes, thereby achieving individual, community and organizational objectives." Kotler (1984, p. 19) offers a more simplified definition in stating that the essence of marketing is "the creation of exchange processes between two or more parties to mutual advantage." Admittedly, these definitions are broad and conceptual in nature, but they indicate that marketing is not a one-way process from seller to consumer/customer, but a continuous two-way (or more) process of activity.

At a World Tourism Organization (WTO) seminar in Ottawa in 1975 a useful tourism-applied definition of marketing was forwarded (Mill and Morrison, 1985, p. 358): "Marketing is a management philosophy which, in light of tourism demand, makes it possible through research, forecasting and selection to place tour-

*For the sake of simplicity, the article refers mainly to the marketing of tourism facilities. The principles and applications illustrated, however, refer equally to recreation facilities.

ism products on the market most in line with the organization's purpose for the greatest benefit." This definition indicates a number of things: that marketing is a way of thinking about a situation that balances the needs of the tourist with the needs of the organization or destination; that tourism marketing research leading to tourist market segmentation is essential; and that the concepts of the product life cycle and development of (new) relevant product offerings are important.

In view of increasing tourism destination competition (e.g., between cities and resorts) during the past ten years, it seems necessary to expand (or to make more explicit) the foregoing definitions of marketing. Marketing today involves not only the creation and offering of appropriate values by an organization whereby mutually beneficial exchanges can be effected, but it is important that it does so better than its competitors. All tourism markets are at any point in time finite in their extent, and heightened competition between tourism destinations for their custom means that destination and product positioning approaches, as part of strategic marketing thinking, should be adopted.

A strategic orientation to marketing is by no means a new concept as the current literature amply demonstrates (see for example: Aaker, 1984; Bonoma, 1985). However, this does not mean that such an orientation has already become widespread in most industries and organizations; on the contrary, it seems that in many industries, and also among non-profit organizations and public authorities, a strategic orientation to marketing, in its contemporary extent, has to be adopted as a matter of urgency (Leibold, 1988).

METHODOLOGY OF RESEARCH

The study made use of both secondary and primary sources of information:

a. Secondary sources of information included relevant published and unpublished material (literature and video material) in both international and domestic (South African) contexts.
b. Primary sources of information involved empirical investigations by way of personal surveys among:

1. Five major city authorities in South Africa with a population of more than 300,000 each
2. A judgemental sample of five local town authorities (town municipal institutions) with populations of less than 60,000 each, in the Western Cape region of South Africa.
3. A relevant spectrum of knowledgeable spokesmen in the tourism industry in South Africa.

The surveys were conducted by way of structured questionnaires. To a considerable degree the data gathered had to be collated and generalized in a manner which was considered to be logical, but could not necessarily be verified statistically. However, various avenues were utilized in order to collect the basic data required, and as such did not only furnish the information, but also acted as checks on one another. Carefully considered evaluations and follow-up contacts were essential in order to come to specific conclusions.

EXISTING APPROACHES TO THE MARKETING OF TOURISM AND RECREATION FACILITIES

Four different approaches to the marketing of tourism and recreation facilities have been determined. These can be described, respectively, as product, selling, consumer needs, and community interest approaches. The nature of each of these approaches are outlined, and then the incidence of use by local authorities thereafter stated.

The Product Approach

A product approach suggests that the emphasis be placed upon the products or services which a destination has available. For example, a destination area has many physical, historical and cultural resources, be they of natural or man-made origin. The orientation inherent in such an approach is that the extent to which one destination considers its resources as "better" than that of competition would determine how many tourists visit their particular destination. The well-known adage emanating from Ralph Waldo Emerson reflects this approach, in stating that "build a better mousetrap, and the world will beat a path to your door" (Kotler, 1984, p. 351).

This orientation is probably rooted in the philosophy of the nineteenth-century economist J. B. Say, who propounded the theory of "supply creates its own demand."

Although it cannot be denied that the quality of resources is important, a total emphasis on tourism supply fails to take the wishes of potential tourists into account. A product orientation may be successful if there is a surplus of demand over supply, but this is not characteristic of our current environment. This orientation has been used by the local authorities of a town on the south coast of England who decided in the late 1960s to print brochures only in English. When it was pointed out that the French residents across the English Channel represented a major potential market, the reply was given that if the French wanted to visit, then they would be interested enough to learn to read English in order to understand what was available (Mill and Morrison, p. 359).

Organizations practicing this approach can be typified as looking in a mirror when they should be looking out of the window. The concentration on the purity and immutability of the product eventually creates considerable marketing problems for such organizations, as some are currently experiencing.

The Selling Approach

The number of destinations actively seeking tourism has increased considerably in the past few years, as has the accessibility to more places by tourists. When there is more supply than demand, it becomes increasingly necessary to sell tourists on the benefits of visiting a particular destination or of purchasing a particular product or service. The orientation for many has shifted from one of emphasizing product to one of emphasizing selling.

The entry into the marketplace of more professional tour packages has increased competition for tourist custom, and this means that destination areas can no longer sit back and wait for tourists to come to them. The emphasis is consequently being placed by many on promotion of what is available for sale. This selling approach focusses on the needs of the seller—which is to sell the product—rather than the needs of the potential consumer.

The implicit premises of the selling concept are that more consumers can be induced to buy through various sales-stimulating devices such as advertising and sales promotions, and that one of an organization's major tasks is to maintain a strong sales-oriented department. There is little research into what consumers want, with the assumption that a destination is "sold, not bought." Obviously, there are great risks in practicing the selling concept, especially in its hard-driving form where customer satisfaction is considered secondary to getting the sale.

The Consumer Needs Approach

A further approach is one in which the needs and wants of the tourist are placed foremost in the mind of the marketer. In short, the consumer needs approach says "find wants and fill them," rather than "create products and sell them." It involves providing the kind of experiences that tourists want rather than what an organization feels that they "should" want.

The underlying premises of the consumer needs approach are:

- Consumers can be grouped into different market segments depending on their needs and wants;
- Consumers in any market segment will favor the offerings of the organization/destination which comes closest to satisfying their particular needs and wants; and
- The organization's task is to research and choose target markets and develop effective offerings and marketing programs as the key to attracting and holding customers.

The selling approach and the consumer needs approach are frequently confused by marketers. Levitt (1960) draws the following contrast between these two orientations: "Selling focusses on the needs of the seller; marketing on the needs of the buyer. Selling is pre-occupied with the seller's need to convert his product into cash; marketing with the idea of satisfying the needs of the customer by means of the product and the whole cluster of things associated with creating, delivering and finally consuming it." In essence, the marketing concept is a consumer needs and wants orientation backed by integrated marketing effort, aimed at generating customer satisfaction as the key to satisfying organizational objectives. In short, it is a commitment to the well-known concept in economic theory known as consumer sovereignty.

The Community Interest Approach

The uniqueness of tourism suggests that a philosophy that concentrates solely on the needs of the market is not an optimal approach. Tourism supply at a particular destination is oriented towards the resources of that community. To become totally consumer needs oriented, all aspects of the community would have to be oriented towards satisfying the needs and wants of the tourist.

The risk for the community, and ultimately for the tourist as well, is that by orienting strictly and totally for the tourist's wants, the needs and integrity of the community may be abused. Cases exist where tourist destination areas that have adapted to the needs of the tourist have, in the process, lost their uniqueness and heritage while receiving a relatively poor economic return on their investment.

An answer to this problem is for a destination to develop a marketing approach that focuses on the satisfaction of tourist needs and wants while respecting the long-term interests of the particular community. The consideration added in the community interest approach is the emphasis on long-run consumer and societal well-being. The underlying premises of this approach are:

- Consumer's wants do not always coincide with their long-run interests or the community's long-run interests; and
- The organization's (or destination's) task is to serve target markets in a way that produces not only want satisfaction but long-run individual and societal benefit as the key to attracting and holding customers.

Although conceptually appealing, major questions facing organizations in the application of this approach are how to optimally reconcile consumer and community interests, and how to react to competitors' behavior without sacrificing long-run objectives. Furthermore, this approach emphasizes community interests, but not all stakeholders' interests. The multi-faceted stakeholders involved, or affected by, the tourism industry of a particular destination necessitates a new, more appropriate approach to the marketing of tourism facilities.

Approaches Used by Local Authorities

The findings concerning the approaches used in the marketing of tourism facilities by local authorities in South Africa can be summarized in two parts, viz. cities and towns.

Cities

Of the five cities surveyed, three (60%) use the selling approach and two (40%) the consumer needs approach. The cities in the latter group have relatively recently (past three years) adopted the consumer needs approach, with activities such as marketing research, market segmentation and market positioning featuring prominently. The three cities using the selling approach emphasize market communication activities such as brochures, sales promotions, advertising and publicity.

Several details of the findings concerning cities are especially noteworthy.

The two cities using the consumer needs approach are attempting to create distinct images of their destination in the minds of their particular target markets. This forms part of corporate and marketing strategy thinking by their tourism marketing organizations, which in one of the two organizations has been made explicit, while in the other it is still of an implicit nature. The major problem noted by both organizations is the relative dearth of tourist market information, and the consequent need for marketing research.

The marketing communication activities practiced by the three "selling-oriented" cities are predominantly aimed at travel agents and tour operators, and tourists already at the destination. It seems logical that such activities should also be aimed more directly at potential tourists in the major tourism markets. The problem is that they have not yet decided upon who their target tourism markets are, which makes it highly likely that a large percentage of their existing expenditure could be ineffective.

The selling approach used is perhaps less one of orientation than of resource constraints. Two of the three cities using this approach mention "that they could do much more" if they were not constrained by limited funds, which also impact on staff and supporting services. This does not, however, alter the fact that they are currently using the selling approach.

Towns

The five towns surveyed all use, to a greater or lesser extent, the selling approach in the marketing of their tourism facilities. The available products are accepted as "given" (or "fixed"), and the town publicity organizations see their brief as one of "making tourists aware of what our town has to offer."

Especially noteworthy among the findings are:

- No marketing research efforts are made, or even attempts to gain basic market information
- There are no planned product development activities
- Selling efforts are not targeted, as predetermination of target markets and objectives are absent
- The unanimous belief is that marketing is identical to selling, and that "marketing" activities is subservient to "tourism development" activities.
- Although all of the five towns are members of a regional tourism body (some with misgivings as to its benefits), they view this body as being a forum where mutual problems can be discussed and regional selling instruments can be created, but not as a facilitator for real coordinated marketing efforts.

In summary, the findings of the study indicate that the selling approach is still the predominant approach used by local authority organizations in the marketing of their tourism facilities. This is not surprising when taking into account that many of the executive staff in such organizations, especially on town authority level, have a town planning or government administration background and not a marketing background. The consequent need for more marketing orientation and thinking is evident, which could be supplied through appropriate training courses and/or suitable personnel.

THE CHALLENGE FOR LOCAL AUTHORITIES: ADOPTING A STRATEGIC STAKEHOLDER APPROACH TO TOURISM AND RECREATION MARKETING

The findings of the study indicated that not only are current orientations toward the marketing concept as such largely deficient, but also that the particular nature of tourism marketing requires a new approach—uniquely suited to tourism complexities—to be developed. The uniqueness of tourism marketing issues is firstly described, and then a new approach to effectively accommodate these issues is proposed.

The Uniqueness of Tourism Marketing

Tourism marketing is unique in three major respects.

Firstly, the service provided is a combination of several products and services—for example, a vacation has a transportation component, an accommodation component, a food and beverage component, an attractions component, an activities component, etc. These components are usually offered by different firms (stakeholders in the destination's tourism industry) and they may be marketed independently of each other or combined into packages whereby one vacation is offered but the services are supplied by the different firms. The lack of overall control of the entire vacation offering to the tourist means that a great deal of interdependence among stakeholders results. The marketing efforts of each of the parts are thus affected by the efforts of the others providing a part of the vacation. To attain a consistent image in the minds of its target markets a tourist destination has to implement some form of consensus about and coordination of the various service offerings in order to provide consistent quality of services to tourists.

A second factor making tourism marketing unique is the role of travel intermediaries. Most tourist destinations are located at considerable distances from their potential customers, which means that specialized intermediaries that operate between the supplier of destination services and the tourist are necessary. While in many other industries the supplier exerts much control over every stage in the communication and delivery of the product, in tourism the travel intermediaries have a direct influence on which services are offered, to whom, when and how, and at what price (Hodgson, 1987, p. 148).

The third factor making tourism different from other industries relates to tourism demand and supply. Tourism demand is highly elastic, seasonal in nature, and subject to factors such as taste, fashion and price. Tourism supply, on the other hand, cannot be stored and is relatively fixed in the short term. The resources and infrastructure of a destination cannot change as quickly as tourist demand; in fact, tourism services have to be planned ahead of time. In many instances the "product" sought by tourists can be satisfied by a number of destinations, which makes the need to create overall competitive advantage for a destination important.

A Strategic Stakeholder Approach to Tourism Marketing

The strategic stakeholder approach to tourism marketing by public authorities involves the viewing of marketing as

1. external environmental behavior to create differential (and sustainable) competitive advantage for the tourism destination;
2. a way to decide upon and optimally combine a number of marketing instruments to be used externally in order to achieve target market objectives;
3. internal environmental and consensus behavior between the stakeholders (i.e., all parties interested in or affecting exchange behavior) of the destination's tourism industry; and
4. a management tool to create or expand the tourism marketing organization of the destination in line with strategy objectives.

Tourism marketing using this approach is typified on the basis of two dimensions: external-internal environmental *behavior* and macro-micro organizational *processes*. The proposed typology is presented in Table 16.1.

The underlying rationale and the four major components of this approach can be elaborated upon as follows.

TABLE 16.1. Components of a Strategic Stakeholder Approach to Tourism and Recreation Marketing

	ORGANISATIONAL PROCESSES	
	MACRO PROCESSES (a)	MICRO PROCESSES (b)
EXTERNAL BEHAVIOUR	. Industry analysis . Corporate strategy formulation . Product portfolio analysis . Growth strategies . Facilitation of (destination) infrastructure development	. Market segmentation . Target market decisions . Marketing instruments - Products & packages - Pricing - Communication - Distribution
	(c)	(d)
INTERNAL BEHAVIOUR	. Stakeholder consensus bodies . Inter-organisational goals . Conflict handling mecha= nisms . Regional co-ordination . Integrated systems . Community education/ goodwill	. Marketing organisation . Organisational objectives and per= formance . Staffing and motivation . Productivity measure= ments . Systems

Conceptual Basis for the Strategic Stakeholder Approach

Most existing marketing approaches are based on the micro economic paradigm (Carmen, 1980). Paradigms may be defined as statements of the proper domain of a science, what questions it should ask, and what rules to follow (Bagozzi, 1976). Paradigms provide models from which particular traditions of scientific research emanate, and which are accepted as ultimate "truths"—in the marketing science, the existing paradigm emphasizes the profitable manipulation of the elements of the marketing mix, as popularized by McCarthy (1960).

A logical expansion of the micro-economic paradigm is the adaptation of marketing knowledge to the areas of non-profit marketing (Kotler, 1982). However, anomalies have arisen for which the "accepted tradition" fails to provide adequate answers. When such anomalies build up and scientists are losing faith, the field enters a crisis stage, culminating in the emergence of a new paradigm (Arndt, 1983). As used in marketing, the macro economic paradigm is undoubtedly valuable in providing normative decision rules for practitioners. However, when it comes to providing a basis for a positive theory of marketing as exchange, structuring of external and internal behavior, and structuring of macro and micro organizational processes, the micro-economic world-view is inadequate.

The strategic stakeholder approach is based upon the political-economy paradigm developed by Arndt (1983). The political-economy paradigm utilizes constructs from organization theory, political science and sociology. Transposing this to marketing, the task of marketing is to enact favorable "exchange ratios" for the variousstakeholder groups—consumers, distributors, suppliers, shareholders, employees, local public, bankers, local and national government, etc.—which give time, money, raw materials and other inputs, and in return receive money, finished products, ego gratification, and other desired benefits. This paradigm is especially relevant for tourism marketing in view of the many stakeholders involved in a destination's tourism industry, and the fact that a destination's tourism organization, with a major function being marketing, having to reconcile the various objectives and benefits desired by its stakeholders (internal marketing) in order to be effective in the marketplace (external marketing).

In discussing the four components of the strategic stakeholder approach as illustrated in Table 16.1, it is important to realize that the components are closely interrelated.

The External Behavior-Macro Processes Component

A tourism marketing organization has to continually strive for an optimal fit between the tourism destination's external environment and its own objectives and offerings—an optimal strategic fit. It has to conduct tourism industry analyses, identifying relevant opportunities and threats, and determining its destination's strengths and weaknesses, in order to define its mission and strategic thrust. Useful tools in this regard are portfolio analyses and policy matrices

(Aaker, 1984), whereby the need for new products can be identified and its development facilitated. Through these activities a destination can ensure optimum value-exchanges with its markets, in a unique or better way relative to competitive efforts.

The External Behavior-Micro Processes Component

To implement its strategy, a tourism organization has to decide upon the particular marketing instruments to be used in particular target markets. In other words, the micro issues of target market decisions, determination of marketing objectives, and decisions concerning the nature, extent and combination of the four marketing "p's"—products, pricing, communication and distribution—have to be considered. It is obvious that these micro issues have a direct and close interrelationship with the macro issues in an external environmental context—for example, target market decisions have a direct bearing on the overall strategic thrust of the tourism destination.

The Internal Behavior-Macro Processes Component

A tourism destination has many stakeholders whose behavior could influence the strategy and operations of its marketing organization, e.g., towns/city council, tourism attraction suppliers, accommodation providers, transportation organizations, tour operators, commercial interests (retailers, restaurateurs, etc.) industrial interests, the general public, regional interests, and non-business groups (environmentalists, conservationists, etc.). The tourism organization has to create stakeholder consensus bodies and inter-organizational goals, implement conflict handling mechanisms, arrange regional coordination of efforts, and fulfill a local public education and information function.

These functions are necessary to ensure unified and concerted internal (destination) behavior, whereby competitive positioning objectives (e.g., particular destination image) can be achieved. The organizational processes in this regard are of a macro nature, as it involves inter-organizational measures.

The Internal Behavior-Micro Processes Component

The internal behavior-micro processes component concerns the intra-organizational (within the tourism organization) arrangements in order to effect consistent and effective internal behavior. Included in this category are decisions about the nature of the marketing organization, its objectives, staffing, productivity and other performance measures, and appropriate systems (such as a planning and control system). The internal behavior in category displays itself in a particular organizational culture and management style.

When viewing all four components simultaneously, it is evident that the effectiveness of the internal behavior of a tourism destination (in macro and micro organizational context) has a direct influence on the effectiveness of its external behavior (again in both macro and micro organizational context). Optimal integration of the issues involved in each component of the strategic stakeholder approach is therefore essential.

CONCLUSION

The nature of the tourism marketing task of local authorities is unique and complex. In an important industry characterized by heightening competition between tourism destinations, the application of existing marketing approaches seems to be inadequate. Furthermore, the considerable number of local tourism industry stakeholders, often with divergent objectives, make internal marketing activities an important function of the tourism organization of a particular local authority.

The strategic stakeholder approach to tourism marketing offers a new conceptual tool to guide marketing efforts. Admittedly, the challenge now arising is to validate this approach through appropriate research methodologies. The constructs and relationships specified are theoretically appealing, but difficult to test through conventional tools such as cross-sectional analysis. However, two investigations by the author based on case studies are in progress, which show promise for scientific conclusions in the near future.

REFERENCES

Aaker, D.A. (1984). *Strategic Market Management.* NY: John Wiley & Sons.

American Marketing Association (1985). *Report of the Committee on Definitions*, A.M.A., Chicago.

Arndt, J. (1983). "The Political Economy Paradigm: Foundation for Theory Building in Marketing," *Journal of Marketing*, Vol. 47, pp. 44-54.

Bagozzi, R.P. (1976). "Science, Politics and the Social Construction of Marketing." In Bernhardt, K.L. (ed.), *Marketing: 1976-1976 and Beyond,* proceedings of the 1976 Educators' Conference, Chicago: American Marketing Association, pp. 586-592.

Bonoma, T.V. (1985). The Marketing Edge: Making Strategies Work, NY: The Free Press.

Bush, R.F. & Hunt, S.D. (1982). *Marketing Theory: Philosophy of Science Perspectives.* Chicago: American Marketing Association.

Carman, J.M. (1980). "Paradigms for Marketing Theory." *Research in Marketing*, Vol. 3, pp. 1-36.

Hodgson, A. (ed.) (1987). *The Travel and Tourism Industry*, Oxford: Pergamon Press.

Kotler, P. (1982). *Marketing for Nonprofit Organizations* (2nd ed.) Englewood Cliffs, NJ: Prentice-Hall, Inc.

______ (1984). *Marketing Management: Analysis, Planning and Control* (5th ed.), Englewood Cliffs, NJ: Prentice-Hall, Inc.

Leibold, M. (1988). Structural Market Changes and Global Retailing Strategy Approaches." In Kaynak, E. (ed.), *Transnational Retailing,* New York/ Berlin: De Gruyter.

Levitt, T. (1960). "Marketing Myopia." *Harvard Business Review*, July-August: pp. 45-56.

McCarthy, E.J. (1960). *Basic Marketing: A Managerial Approach*, (7th ed.), Homewood, IL: Richard D Irwin.

Mill, R.C. & Morrison, A.M. (1985). *The Tourism System: an Introductory Text*, Englewood Cliffs, NJ: Prentice-Hall, Inc.

Chapter 17

The Possibilities of Foreign Tourism Market Segmentation and the Evaluation of the Tourism Supply of SR Croatia (Yugoslavia)

Sandra Weber
Neda Telišman-Košuta

INTRODUCTION

In what way do foreign tourists perceive a certain tourist destination? How do different market segments assess the elements of the tourism supply? Which market segments compose the foreign tourist demand? What are the implications of differences between market segments for the creation of a tourism product and for its promotion? The list of questions a tourism planner can ask in order to assure a planned development of a tourist locality is practically endless.

On the other hand, behavior of the foreign tourist demand is very dynamic and liable to constant change. It is impossible to define "uniform and constant" consumer segments because the motives for travel, attitudes toward certain destinations, tourists activities, demographic and psychographic characteristics of potential consumers change with time. Therefore, tourism market research and segmentation, which are the basis and the prerequisites of a market orientation, are also a continuous task.

In an endeavor to satisfy such research demands, the Institute for Tourism, Zagreb and the Committee for Tourism of SR Croatia carried out, from June 15 to September 15, 1987, a survey entitled

"Attitudes and Expenditures of Foreign Tourists–TOMAS '87."[1] The principal goal of this survey was to examine the main features of the foreign tourists demand in SR Croatia[2] such as its socio-demographic characteristics, to provide information related to the tourists' travel and stay in a destination, to determine their expenditures and their attitudes toward and evaluation of the tourism supply elements.

More precisely, TOMAS '87[3] provides the following information about the respondents:

– country of permanent residence, sex, age and profession
– means of transportation
– sources of information consulted before vacation
– type of trip and motives for coming
– individual or organized trip
– reservation of accommodation and arrangements for food
– previous visits and length of stay
– intention to come back again in the next two or three years
– overall evaluation of the 23 tourism supply elements in destination (satisfaction ratings 1-4 and importance of features)
– expenditures in a destination.

Application of such data about the market makes it possible to evaluate the tourism supply elements and it provides possibilities for tourism market segmentation according to various characteristics which, in turn, allows a tourism professional to determine the most suitable product types, their promotion and distribution.

METHODOLOGY

The TOMAS '87 survey was carried out among foreign tourists in 42 resorts of SR Croatia during the main tourist season of 1987 (June 15-September 15). A total number of 14,160 questionnaires were distributed with a response rate of 59%.

The questionnaire (printed in four languages) was used as the main research instrument. It was distributed to a randomly selected tourist sample in preselected hotels of all categories, in tourists' settlements, private rooms or apartments, campsights and motels.

The tourists were asked to fill in the questionnaires themselves.

The principal data about the foreign tourist demand was obtained through basic data analysis methods: frequencies and contingency tables. In addition, factor and discriminant analyses were used with the aim of discerning the existing market segments.

The factor analysis is usually used to determine whether there is a connection between variables, thus making it possible to reduce a great number of variables into few variable groups. The discriminant analysis identifies variables which are the best predictors of belonging to certain groups; it explains differences between chosen groups and provides functions which differentiate them the best. Both of these multivariate methods were used to analyze the respondents' evaluation of the Croatian tourism supply elements.

RESULTS AND ANALYSIS

The Possibilities of Foreign Tourism Market Segmentation in SR Croatia

The entire tourism market, defined as a set of all actual or potential buyers, is too broad and diverse in its needs for any one destination to satisfy them with its particular travel product. Therefore, the processes of market segmentation and target marketing require the following steps:

1. Identifying and distinguishing between different groups/segments which make up the market
2. Choosing one or more segments to focus on
3. Developing marketing strategies (products, price, place, promotion) to meet the needs of the selected target market(s).

Market segmentation is the process of dividing a total market into distinct and meaningful groups of people who have relatively similar needs, interests and wants. The main rational reason for segmenting the market is the possibility for a tourist organization or region to develop a marketing strategy that satisfies a segment of the total, heterogeneous market, instead of attempting to satisfy the needs of all tourists. The goal of market segmentation is to divide a market in such a way that within each segment the individuals respond similarly to a given marketing program. Market segmentation helps in identifying those market segments which would be most readily satisfied by the offered product. The starting point for analysis of the potential target market is knowledge of present customers and their main characteristics. To be effective, market segmentation has to satisfy the following conditions:

1. The travelers' needs for a product or service must be different. If the needs do not differ, there is no point in segmenting the market. The tourism market is characterized by many variations in their need for the travel product.
2. The segments must be identifiable and divisible. There has to be a basis for effectively separating individuals in the market into segments that have relatively similar needs.
3. The total market should be divided in a way which enables the estimation of each segment's buying potential.
4. At least one segment must have enough profit potential to justify the development of a special marketing strategy for that segment.
5. It must be possible to reach the chosen segment and to influence it with a particular marketing strategy.

The tourism market can be segmented in a number of different ways.[4] Kotler's diagram "Steps in Market Segmentation, Targeting and Positioning"[5] served as the basis for Figure 17.1 which shows how data on foreign tourists, gathered in this survey, can be used in market segmentation. Identifying tourists by demographic variables is usually the first step in distinguishing customer groups. The next step is understanding their needs, motives for travel and their behavior, which are used in psychographic[6] and benefit segmentation.[7] In addition to socio-demographic characteristics, the TOMAS '87 survey results offer data about tourists' attitudes toward the destination assets and facilities, evaluation of supply elements and the importance of these elements, all of which are inputs for market segmentation.

Thus, it is possible to segment the tourism market of SR Croatia combining the respondents' attitudes toward the 23 tourism supply elements[8] with their demographic characteristics. This manner of

FIGURE 17.1. Market Segmentation of SR Croatia

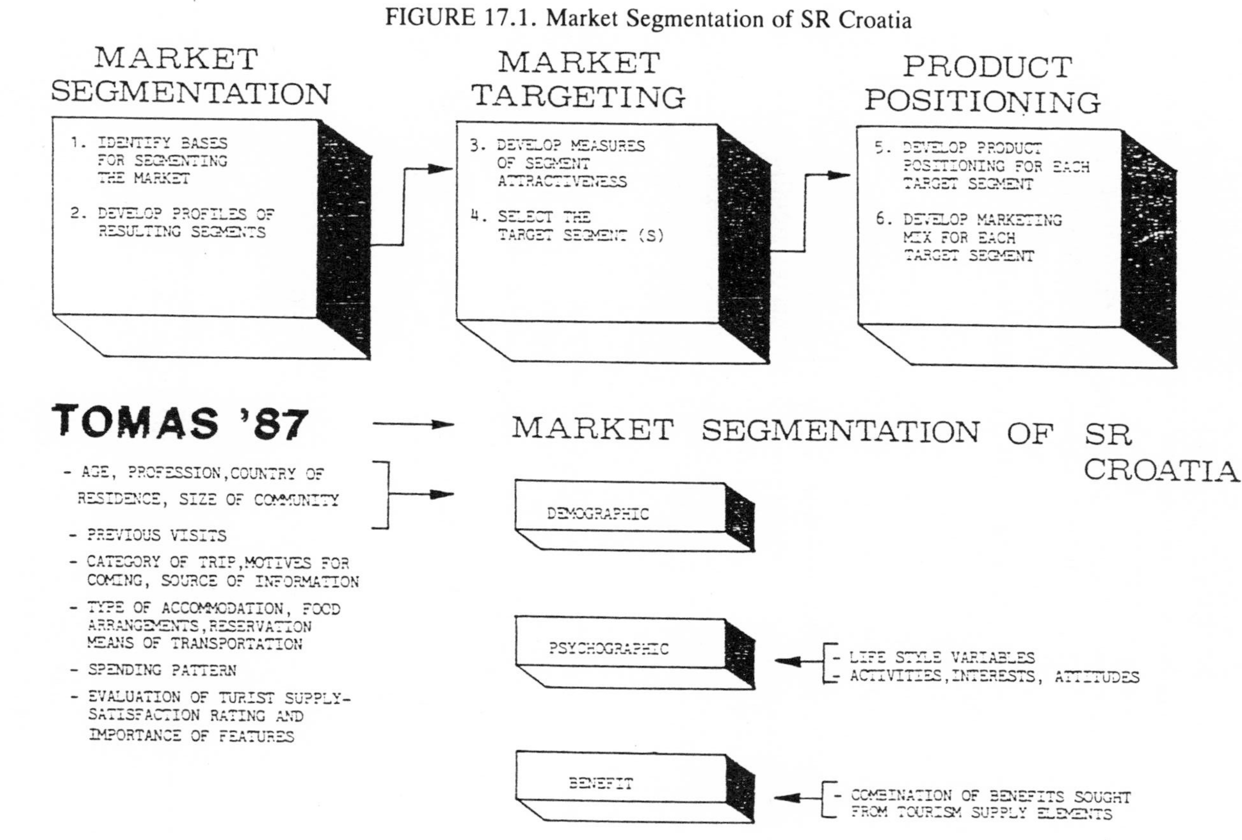

combining demographic segmentation with segmentation based on attitudes toward the tourism supply will enable the easiest communication process with the market and make product development and positioning possible.

The main task of segmenting methods is to, first of all, discover groups of similar needs and motives, and then to determine differences between them and to pinpoint the distinguishing features. These were also the main tasks in segmenting the foreign tourist market in this study.

Grouping the Respondents According to Their Attitudes Toward the Tourism Supply

The starting assumption underlying the analysis was that the respondents perceive different supply elements differently, or in other words, that certain supply elements are related to each other to a greater extent than they are connected to some other element group, thus forming a specific product type.

The results of the factor analysis confirmed this assumption because the respondents' answers formed 6 groups or factors explaining 53.1% of the total variance. However, the structure of each factor showed that, assuming the arbitrary correlation coefficient of 0.48, only 4 factors are interpretable, explaining 40% of the total variance (Table 17.1).

The four interpretable factors show that the respondents' answers are grouped around certain supply elements, which may be labeled as follows:

FACTOR 1 – ENTERTAINMENT, SPORTS, SHOPPING AND OTHER ACTIVITIES

– entertainment facilities
– range of sports facilities
– facilities for conferences, meetings, etc.
– walking and hiking paths
– scope of excursions
– cultural life
– shopping facilities
– suitability for family vacations.

FACTOR 2 – BASIC SUPPLY ELEMENTS

– service in the place of accommodation
– comfort in the place of accommodation
– standard of catering in the place of accommodation
– value for money.

FACTOR 3 – ENVIRONMENTAL ELEMENTS

– picturesquesness of town/village
– environment and countryside
– peace and quiet
– walking and hiking paths.

FACTOR 4 – SUITABILITY FOR FAMILY VACATION

– friendliness of local people
– suitability for family vacation
– safety while staying in a destination
– information provided by local tourist or information office
– shopping facilities
– climate and weather.

The most obvious fact in the case of Factor 1, *Entertainment, sports, shopping and other activities* is its heterogeneous structure. It could be concluded that for creating the tourism product it is equally important to consider cultural, sports and all other accompanying services, representing diverse fields of human activity, as parts of the tourism product "amalgam."[9] Therefore, all the included services and activities need to be treated as a specific tourism industry sector which is not based on the usual "touristic" attributes, but on socio-economically distinguished elements, functioning as a receptive tourism community.

Input variables of Factor 2, *Basic supply elements*, are logically inter-connected. The factor groups variables that represent certain basic dimensions of the tourism supply forming the foundation of tourism as a service industry. The inclusion of "value for money" in this factor indicates that this group of variables is perceived

TABLE 17.1. Factor Analysis – Correlation Coefficients of the Components

COMPONENTS:	FACTOR 1	FACTOR 2	FACTOR 3	FACTOR 4
1. ENVIRONMENT AND COUNTRYSIDE	0,242	0,047	0,704	0,246
2. CLIMATE AND WEATHER	0,211	0,167	0,091	0,476
3. SAFETY WHILE STAYING IN A DESTINATION	0,220	0,318	0,414	0,495
4. PICTURESQUENESS OF TOWN/VILLAGE	0,392	0,252	0,746	0,234
5. PEACE AND QUIET	0,170	0,265	0,602	0,157
6. ACCESSIBILITY	0,381	0,217	0,155	0,270
7. LOCAL TRAFFIC CONDITIONS	0,327	0,200	0,316	0,215
8. COMFORT IN THE PLACE OF ACCOMM.	0,383	0,810	0,312	0,290
9. SERVICE IN THE PLACE OF ACCOMM.	0,349	0,840	0,209	0,412
10. STANDARD OF CATERING IN THE PLACE OF ACCOMM.	0,402	0,730	0,179	0,227
11. SERVICE IN RESTAURANTS	0,312	0,289	0,206	0,305
12. STANDARD OF CATERING IN RESTAURANTS	0,327	0,243	0,174	0,275
13. FRIENDLINESS	0,318	0,335	0,296	0,752
14. CULTURAL LIFE	0,628	0,244	0,402	0,401
15. INFORMATION PROVIDED BY LOCAL TOURIST OFFICE	0,545	0,328	0,313	0,487
16. VALUE FOR MONEY	0,439	0,509	0,364	0,316
17. EXCURSIONS	0,632	0,237	0,341	0,155
18. WALKING AND HIKING PATHS	0,641	0,176	0,479	0,150
19. ENTERTAINMENT	0,700	0,205	0,180	0,365
20. FACILITIES FOR CONFERENC.	0,641	0,279	0,240	0,251
21. SHOPPING FACILITIES	0,570	0,196	0,163	0,484
22. SUITABILITY FOR FAMILY VACATION	0,521	0,334	0,221	0,687
23. SPORTS FACILITIES	0,669	0,154	0,141	0,334
% VARIANCE	15,5	9,2	7,8	7,5

through the received value for money given. This is the most important reason to keep this part of the tourism supply competitive.

Factor 3, *Environmental elements*, also represents additional important dimensions of the tourism supply. Some input variables of this factor, such as the "environment and countryside," were highly ranked on the list of all evaluated supply elements, but others, such as "walking and hiking paths," were evaluated rather poorly. This indicates that it is not possible to set the resource base apart from the way it is managed. In other words, it is obvious that the environment represents a resource of the tourism industry only if facilities, such as walking and hiking paths, are available to enjoy it.

Factor 4, *Suitability for family vacation*, groups various variables emphasizing the interaction between guests and hosts ("friendliness," "information"). This factor also includes "safety while staying in a destination," showing that this element is an important part of this perceived type of tourism product.

The factor analysis has, based on consumer attitudes, grouped particular tourism supply elements, at the same time showing both the respondents' perception of the destination features[10] and pointing to the specific types of tourism supply. In this phase of the research, then, the grouping of respondents' perceptions has indicated the potential "types of tourist products."

Differentiating Respondent Groups According to Their Attitudes Toward the Tourism Supply

Since the principal goal of market segmentation is the grouping of customers who are in their needs and motives homogeneous within the group and heterogeneous in regard to other groups, the discriminant analysis[11] was judged as the appropriate statistical method to be employed.

For applying this multivariable analysis, groups must be identified a priori. Therefore, in order to analyze the respondent's evaluation of the 23 supply elements, the following respondent characteristics were used to form the discriminating groups:

– age,
– profession,
– size of the community, and
– country of permanent residence.

In this work, only the results of the discriminant analysis distinguishing *age groups* and *groups according to the country of permanent residence*, will be presented, since these were found to be the most discriminative features. See Tables 17.2 and 17.3, Figures 17.2 and 17.3.

Discriminant function coefficients show that all three age groups are distinguished from each other in the case of both discriminant functions: *Entertainment and suitability for family vacation, and Most important supply elements.* In the case of the *Most important supply elements*, the furthest distance between centroids occurs between the oldest and the youngest respondent groups while those aged 30-49 are closer to the youngest. It can be concluded that the

TABLE 17.2. Age of Tourists

	CANONICAL CORRELATION	EIGEN-VALUE	CHI-SQUARED	DEGREE OF FREEDOM	SIGNIFI-CANCE	WILK'S LAMBDA	% VARIANCE
1.	0,297	0,097	202	46	0,0000	0,89	80,90
2.	0,149	0,023	40	22	0,0119	0,98	19,10

DISCRIMINANT VARIABLES (COEFFICIENTS)

GROUP CENTROIDS	I	II
I. UP TO 29 YEARS (n=572)	-0,319	0,145
II. 30 - 49 YEARS (n=798)	-0,019	-0,165
III. 50 AND ABOVE (n=400)	0,494	0,121

FIGURE 17.2. Age of Tourists (Years)

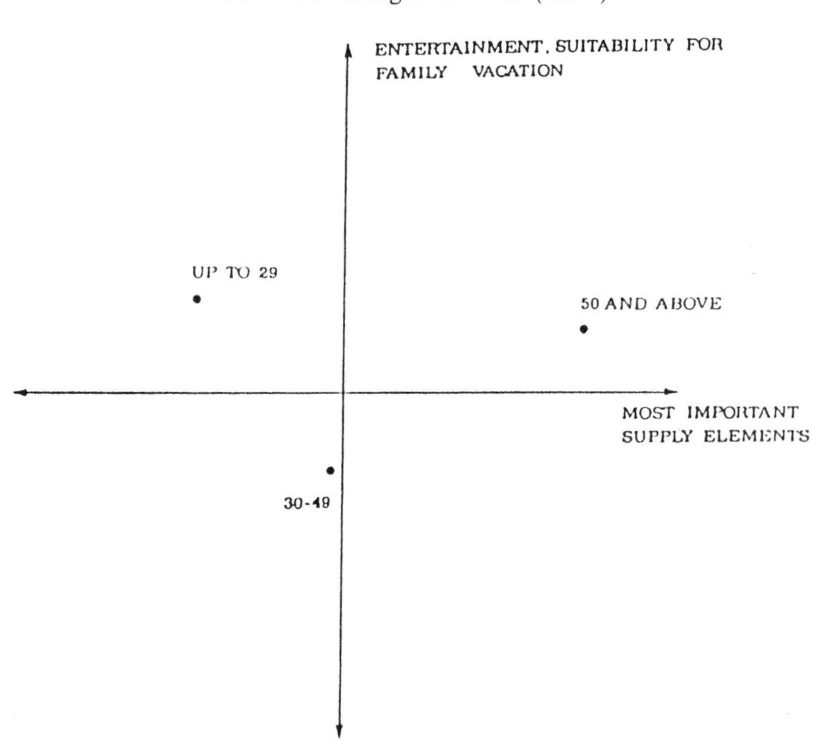

TABLE 17.3. Country of Residence

	CANONICAL CORRELATION	EIGEN - VALUE	CHI-SQUARED	DEGREE OF FREEDOM	SIGNIFI-CANCE	WILK'S LAMBDA	% VARIANCE
1.	0,454	0,260	871	92	0,0000	0,61	48,32
2.	0,400	0,190	461	66	0,0000	0,77	35,45

GROUP CENTROIDS	DISCRIMINANT VARIABLES (COEFFICIENTS) I	II
I. FR GERMANY (n=716)	0,201	0,345
II. GREAT BRITAIN (n=511)	-0,700	-0,137
III. AUSTRIA (n=139)	0,271	0,369
IV. NETHERLANDS (n=141)	0,233	-0,045
V. ITALY (n=285)	0,502	-0,778

FIGURE 17.3. Country of Residence

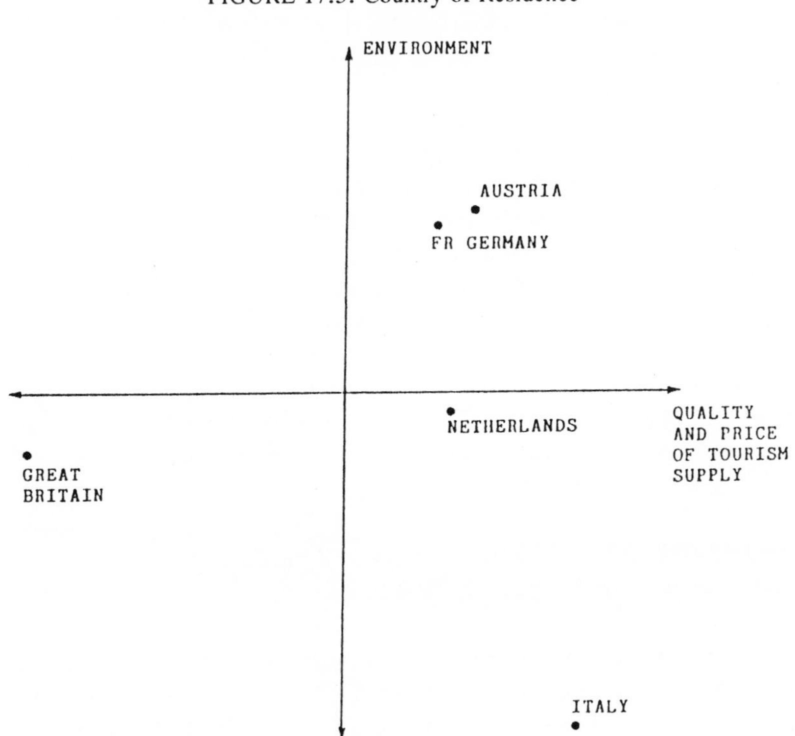

three age groups have different perceptions and attitudes toward the *Most important supply elements*. In the case of the second variable—*Entertainment and suitability for family vacation*, the youngest tourists and tourists between 30 and 49 years differ the most, while, in this case, the oldest and the youngest show similar attitudes.

It can be concluded that *age* is a variable which greatly distinguishes tourists in relation to their opinions and attitudes. Tourists over 50 could be treated separately from the two younger groups. This has to be taken into account in creating and developing the tourism product.

In reference to the tourists' country of permanent residence (Figure 17.3), the position of centroids on the first and the most important discriminant function, "*Quality and price of tourism supply*," indicates the greatest perceptual differences occur between tourists from Great Britain and those from Italy. Respondents from Austria, FR Germany and the Netherlands are positioned in-between them, showing their similarities toward quality and price aspects of the tourism supply.

An examination of the manner in which the various nationalities actually rated the discussed supply elements reveals that the Italians rated them rather highly while the tourists from Great Britain were the most critical. Guests from FR Germany and the Dutch were more inclined toward middle-of-the-road marks.

Considering the location of group centroids it could be concluded that the demographic characteristics *age* and *country of residence* significantly distinguish groups of respondents and are very appropriate variables for segmenting the foreign tourism market of SR Croatia. The main conditions of market segmentation—combining tourists into particular groups, while distinguishing them at the same time, have also been met.

The Possibilities of Tourism Supply Evaluation with Respect to Target Market Segments

Each of the discussed analyses provided a basis for selecting target market segments on which the marketing strategy needs to focus. Of course, first it is necessary to choose the criteria for target group selection, and then make a decision to target certain markets. Since this is a task within the domain of a certain region's or organization's tourism policy, in this article only the characteristics and the possibilities for selection of certain segments will be pointed out. The most frequently used criteria for the selection of target groups include the following:[12]

1. size of the market segment;
2. possibility of locating, identifying and communicating with the segment;
3. tourist expenditure; and
4. degree to which a certain region's tourism supply is adapted to the needs and the demands of the market segments.

After defining the "target markets" according to each of the above criteria, and taking into account conditions within the general economic environment, it is possible, for a specific time period to determine the target groups, which the marketing strategy will address. The marketing strategy will, then, define the elements of the marketing mix—product, price, promotion and distribution—suited to the chosen target segments.

The TOMAS '87 results allow a grouping and differentiating between consumer groups according to their socio-demographic characteristics and their attitudes toward the tourism supply, as well as making make it possible to evaluate the tourism supply. In Figure 17.4 the elements of the Croatian tourism supply are evaluated from the consumer's point of view, or in other words, the elements are grouped into what the tourists perceive as certain "tourism products": *entertainment, sports, shopping and other activities*, *basic supply elements*, *environmental elements*, and the, so called, *suitability for family vacations elements*. The elements enclosed within frames are those supply elements which the tourists have evaluated either as the supply's "strategic success positions" (the elements evaluated as most favorable and, at the same time, considered most important) or as its "goodwill positions" (ranked among the most favorably evaluated elements, but not considered as the most important). Thus, these are the supply elements which represent the positive aspects of the Croatian tourism supply.

FIGURE 17.4. Evaluation of the Tourism Product—Consumer Oriented

ADVANTAGES FOR: CHOSEN TARGET MARKET (S)

DISADVANTAGES FOR: CHOSEN TARGET MARKET (S)

ENTERTAINMENT, SPORTS, SHOPPING AND OTHER ACTIVITIES

CHOICE OF EXCURSIONS*)
SUITABILITY FOR FAMILY VACATIONS
SPORT
ENTERTAINMENT FACILITIES
CULTURAL LIFE
WALKING/HIKING PATHS
FACILITIES FOR CONFERENCES
SHOPPING FACILITIES

BASIC SUPPLY ELEMENTS

SERVICE IN PLACE OF ACCOMMODATION
COMFORT IN PLACE OF ACCOMMODATION
STANDARD OF CATERING IN PLACE OF ACCOMMODATION
VALUE FOR MONEY

ENVIRONMENTAL ELEMENTS

PICTURESQUESNESS OF TOWN/VILLAGE
ENVIRONMENT AND COUNTRYSIDE
WALKING/HIKING PATHS

SUITABILITY FOR FAMILY VACATION ELEMENTS

CLIMATE AND WEATHER
SECURITY
SUITABILITY FOR FAMILY VACATIONS
FRIENDLINESS OF THE LOCAL PEOPLE
INFORMATION/TOURIST OFFICES
SHOPPING FACILITIES

*) Framed elements represent the "strategic success positions" and the "goodwill positions".

Supplementing the results of the factor and the discriminant analyses with the percentage of respondents rating each supply element as either "quite good" or "very good," it has become possible to segment the foreign tourist market according to the, so called, preferred *product types*. As an example, the *environmental elements* (Table 17.4), and those making up the group of elements labeled *suitability for family vacations* (Table 17.5), which are evaluated as the most favorable, will be analyzed.

Two of the elements comprising the *environmental* factor were rated very highly by the respondents ("environment and countryside" ranked in 2nd place, and "picturesquesness of town/village" in 5th from a total of 23 evaluated elements), while the remaining two were not ranked among the top nine elements. Such a result implies that this supply segment, which in a comprehensive marketing conception could be called the *environmental tourism product*, is not a complete one since the foreign tourists' evaluation of its different components varies. In respect to market differentiation, it is readily apparent that the attitudes toward all the supply elements, with the exception of "walking/hiking paths," can be segmented according to various criteria. If one looks at the first variable among the environmental elements, "the picturesquesness of town/village," it can be seen there is a clear polarization of respondents' attitudes between the oldest (50 and above) and the youngest (up to

TABLE 17.4. Environmental Elements

ENVIRONMENTAL ELEMENTS	PERCENTAGE (quite good+very good)	DISTANCE BETWEEN RESPONDENT GROUPS DISCRIMINANT ANALYSIS			
- PICTURESQUESNESS OF TOWN/VILLAGE	89,2%	-50 and above -housewifes, retired -Italy	-self-employed, executives -Austria, Netherlands, FR Germany	-30-49 years -salaried emply. workers -Great Britain	-up to 29 -students
- ENVIRONMENT	94,9%	-50 and above -housewifes, retired -Austria, FR Germany	-self-employed, executives	-30-49 years -salaried emply. workers -Great Britain, Netherlands	-up to 29 -students -Italy
- PEACE AND QUIET	71,6%	-housewifes, retired -above 100.000 inhabitants -Austria, FR Germany	-self-employed, executives	-salaried emply. workers -Great Britain, Netherlands	-students, -up to 100.000 inhabitants -Italy
- WALKING/HIKING PATHS	69,5%	-50 and above		-30-49 years	-up to 29

TABLE 17.5. Suitability for Family Vacation Elements

SUITABILITY FOR FAMILY VACATION ELEMENTS	PERCENTAGE (quite good+very good)	DISTANCE BETWEEN RESPONDENT GROUPS DISCRIMINANT ANALYSIS			
- FRIENDLINESS OF LOCAL PEOPLE	86,0%	-50 and above -housewifes, retired	-self-employed, executives	-30-49 years -salaried emply. workers	-up to 29 -students
- SUITABILITY FOR FAMILY VACATION	88,8%	-up to 29		-50 and above	-30-49 years
- SAFETY	90,3%	-50 and above -housewifes, retired -Italy	-self-employed, executives	-salaried emply. workers	-students
- INFORMATION/TOURIST OFFICES	83,4%	-housewifes, retired	-self-employed, executives	-salaried emply. workers	-students
- SHOPPING FACILITIES	56,1%	-housewifes, retired	-self-employed, executives	-salaried emply. workers	-students
- CLIMATE AND WEATHER	97,3%	-up to 39 -over 100.000 inhabitants -Italy		-50 and above -Austria, Netherlands, FR Germany	-30-49 years -up to 100.000 inhabitants -Great Britain

29) groups, between the housewives and the retired on the one hand and the students on the other, and between the Italians and the British. The most similar views are held by the two middle segments, consisting of executives, managers, salaried employees, skilled workers, those 30 to 49 years old, from Austria, the Netherlands and FR Germany, who also comprise the largest segment of the foreign tourist demand in SR Croatia. In the case of other variables constituting this factor (except "walking/hiking paths" as it was already mentioned), the market is differentiated in pretty much the same manner. Of course, when considering the *environmental tourism product*, it almost goes without saying the aim of market segmentation is not a question of adjusting nature to the demands' special requests, but it is a question of adjusting those elements which make a tourism resource out of nature.

The elements which jointly form the *suitability for family vacations* factor are the most favorably perceived segment of the Croatian tourism supply. Out of its six variables, the foreign tourists rated four as very good ones. Thus, with further improvements of the tourist information network, and particularly of the possibilities for shopping, it may be possible to create an integral tourism product which would be tailored to the market demands. Here, it is also necessary to take into account the fact that the market is segmented. The discriminant analysis has shown polarized perceptions, in the case of "climate" for example, between Italians and the British, between large and small city dwellers and between those up to 29 years of age and those from 30 to 49. Finally, it is necessary to point out that these results were obtained from a sample spread out over entire SR Croatia, and they must, therefore, be thought of as *global indicators*, while each individual destination has its own specific qualities which dictate the formation of its tourism supply.

Using information about foreign tourism markets in this manner has, in addition to market segmentation, made it possible to evaluate the tourism supply and draw certain conclusions regarding the advantages and the disadvantages of certain supply elements in respect to selected target segments. Based on such research results, the state or regional tourism organizations, when forming their tourism product and drawing up the promotional strategy, can differentiate their presentation on the foreign market according to the selected target group's demands and wishes.

CONCLUSION

Information about the tourist market is the starting point for defining elements of a marketing strategy—tourism product, distribution, price and promotion—aimed at the target segments. The principal goal of the TOMAS '87 survey was to examine the main behavior characteristics of foreign tourists coming to SR Croatia from the most important generating countries, and, using that information as input data, to identify the basis of market segmentation, and to evaluate the tourism supply from the target market's point of view.

The maturity of tourism as a product, or, in other words, its stage in the product life-cycle, has brought about the "end of differentiation and the beginning of the segmentation era."[13] Because of an increasingly competitive international market, it is necessary to adapt to an ever more sophisticated consumer who is demanding, quality-oriented, better-educated and highly informed. Success in the market is highly dependent on a marketing mix which focuses on certain consumer segments, while the tourism product being offered by SR Croatia, today, is a reflection of an unconscious and incidental product development.

Based on the respondents' evaluation of the Croatian tourism supply, four "tourism products" can be defined—*entertainment, sports, shopping and other activities*, then, the *basic supply elements*, *environment*, and, finally, the *family vacation product*. The respondents have indicated the first two of these "products" are significantly less adjusted to the consumers' demands than the *environment* and the *family vacation* which were perceived as the most satisfactory. Such a manner of market differentiation and the socio-demographic consumer segments as inputs for a discriminant analysis produces results which will enable the tourism industry to formulate a tourism product based on consumer demands and activities and adapted to selected target segments.

For successful participation in the international tourist market Yugoslavia has to offer to the potential consumers a competitive product, adapted to their tastes and preferences. A tourist product with an image composed of natural beauty, friendliness of the people and reasonable prices is offered by a whole range of countries. Therefore, through market segmentation it is possible to develop products and marketing strategies to meet the needs of the selected target market(s).

Through a continuous monitoring and evaluation of tourist behavior, and respecting changes in certain markets, the first condition of successful operation and management—operating on the basis of information about the market—will be met.

NOTES

1. TOMAS is abbreviation of *Touristiches Marktforschungssystem Schweiz* (Touristic Market Research System Switzerland), a methodology introduced by Switzerland in winter season 1982/83, followed by Austria (summer 1984) and Yugoslavia (summer 1987).

2. Croatia is one of six socialist republics in Yugoslavia with a population of 4.7 million covering an area of 56.538 km2 and occupies 95% of the Yugoslav coast line. In 1987 Croatia achieved 68.1 million nights spent by tourists with a relative share of 62% of foreign tourists thus absorbing 81% of foreign tourists nights in Yugoslavia.

3. Group of authors: "Attitudes and Expenditures of Foreign Tourists—TOMAS '87," I and II report, Institute for Tourism, Zagreb, Yugoslavia, 1988.

4. Bryant, Barbara E. & Morrison, Andrew J. (1980). "Travel Market Segmentation and the Implementation of Market Strategies," *Journal of Travel Research*, Volume XVIII, Number 3, Winter, pp. 2-8.

5. Kotler, Philip. (1986). *Principles of Marketing*, Englewood Cliffs, NJ: Prentice Hall, p. 263.

6. Ritchie, J.R. Brent & Goeldner, Charles, R. (1987). "*Travel, Tourism and Hospitality Research*." NY: John Wiley and Sons, Inc.

7. Woodside, Arch & Jacobs, Laurence W. (1985). "Step Two in Benefit Segmentation: Learning the Benefits Realized by Major Travel Markets." *Journal of Travel Research,* Summer, Volume XXIV, Number 1, pp. 7-13.

8. The 23 evaluated supply elements include the following: environment and countryside, climate and weather, safety while staying in a destination, picturesqueness of town/village, peace and quiet, accessibility, local traffic conditions, comfort in the place of accommodation, service in the place of accommodation, standard of catering in the place of accommodation, service in restaurants, standard of catering in restaurants, friendliness of the local people, cultural life, information provided by local tourist or information office, value for money, choice of excursions, walking and hiking paths, entertainment facilities, facilities for conferences, shopping facilities, suitability for family vacation, range of sports facilities.

9. Medlick, S. & Middleton, V.T.C. (1973). "The Tourist Product and Its Marketing Implications," *International Tourism Quarterly*, Number 3, pp. 28-35.

10. Calantone, Roger J. & Johar, Jotindar S. (1984). "Seasonal Segmentation of the Tourism Market Using a Benefit Segmentation Framework." *Journal of Travel Research*, Fall, Volume XXIII, Number 2, pp. 14-24.

11. Pitts, Robert, E. & Woodside, Arch, G. (1986). "Personal Values and Travel Decisions," *Journal of Travel Research*, Volume XXV, Summer, Number 1, pp. 20-25.

12. *Creating Economic Growth and Jobs Through Travel and Tourism,* U.S. Department of Commerce, Economic Development Administration, Washington, USA, February 1981.

13. Pizam, Abraham (1987). "Macrotrends in U.S. Tourism and their Impact on U.S. Tourism to Yugoslavia." Report presented in Belgrade.

Chapter 18

Marketing of Tourism in a Competitive World Environment

Ugur Yucelt
Phyllis W. Isley

INTRODUCTION

The cultivation of a strong tourist industry has become a substantial element in the development strategy of a number of countries, in particular, because the character of the activity makes promising social as well as economic contributions to the host country. A labor-intensive service industry, tourism tends to generate employment across a broad spectrum of job skills and levels of education, directly creating jobs in management, maintenance, and for artisans and craftsmen. Secondarily tourism affects employment in other sectors of the economy such as agriculture, food processing, and light manufacturing, as well as engenders support for traditional handicraft industries. In addition to the very positive effects it has on employment, a strong tourist industry can provide an important source of foreign exchange.

A major problem in founding a tourist industry is the large initial investment required. For this reason, even in the industrialized developing nations, the establishment of a tourist industry is typically a joint venture between the host country and multinational companies specializing in tourism. These companies provide not only the initial capital investment in facilities, but also training for management and support staff. They provide essential marketing expertise as well and a marketing network that would otherwise be unavailable to the host country.

The scale of tourism development, on the one hand, is related to operational and cost efficiency of tourism products; location, quality, and level of enclave development; and marketing strategies and tactics. On the other hand, tourism helps to increase job opportunities, entrepreneurship, foreign exchange and investment opportunities, and gross national products. The effective strategies for tourism marketing, therefore, must combine the elements of market structure, economic condition, life cycle of tourist products, and visitors' income as it is presented in Figure 18.1.

To the extent that a joint venture is an essential requirement of success, the host country must provide a cohesive plan which ac-

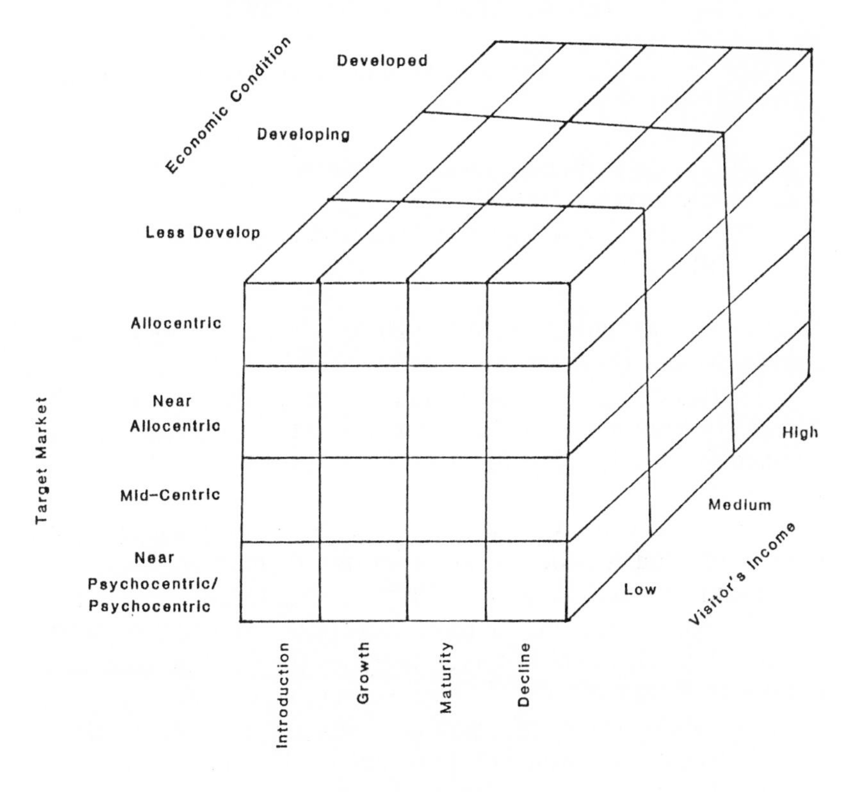

FIGURE 18.1. Strategic Model for the Tourism Industry

commodates the needs of major development projects: land acquisitions, and the cultivation among the populace of a hospitable attitude. Both as information for internal planning for optimum infrastructural development and as a way to attract joint ventures, estimates of the type of demand and the expected level of demand of tourist services should be a strategic element of the development program. This chapter focuses on a forecasting methodology that can be useful for estimating the level of expected demand and suggests strategic alternatives for tourism planning in a competitive environment.

TOURISM AS AN INDUSTRY

A host country in today's world tourist market can attract three different types of foreign visitors:

1. high income, high social status visitors
2. middle income visitors
3. low income, low social status visitors. (Jenkins 1982; Kadt 1979)

High income and middle income visitors are a major source of foreign exchange. These visitors usually seek a location that provides a social structure and environmental amenities that resemble their customary environmental characteristics. If the host country has not attained this level of development, it can usually attract these customers only by providing enclaves and entire resort complexes which offer exclusivity. This is the most difficult route to the development of a tourist industry, since it often alienates the local population and produces fewer employment benefits, both direct and indirect. For this reason, the more successful host countries tend to target the development of tourist facilities for those geographic areas where there have already been substantial economic gains and for those areas where the resort complex supplements an already well-developed infrastructure. Low income, low status visitors, like college students, place less emphasis on quality facilities. They are more casual, less demanding travelers, but they also generate less foreign exchanges.

The tourist market, in addition, is broken down into four segments: (1) allocentrics, (2) near allocentrics, (3) mid-centrics, and (4) near-psychocentrics and psychocentrics (Reime and Hawkins, 1979). The allocentric segment is very small (3-5%) and seeks a culture and environment different from his/her own. Members of this group enjoy a high income level and travel more and farther than the members of most other segments. However, because of its limited size, the allocentric segment is not a dependable source for tourism growth.

The near-allocentric segment is larger (10-15%) and possesses many of the same characteristics as the first category. However, these tourists, while more outgoing, like to feel "at home" in a foreign country and look for luxury and security during their travels. If the infrastructure and superstructure are sufficient, the near-allocentric tourists will contribute a great deal to overall tourism growth.

The mid-centric segment is the largest (60-65%) and tends to share the opinions of the near-allocentric segment. Members of this group generally select a destination similar to their own environment, with minor cultural differences. In order to satisfy the needs of this group, a host country must develop social activities, historical sites, recreational opportunities, shopping areas, an above-average infrastructure, and food and lodging facilities. In order to attract this segment and become competitive, the host country should continually seek to upgrade its services to the expectations of mid-centric tourists.

Members of the near-psychocentric and psychocentric segment, the remaining segment (15-20%), prefer a destination that does not exhibit its foreign culture. These tourists spend all of their time inside tourism complexes, stay for shorter periods of time, and are not big spenders. They are bargain hunters, and look for low-cost vacation spots.

Attractiveness of a host country to different segments, accordingly, is related to a number of important factors: (1) price level of tourism products; (2) infrastructure of the region; (3) natural beauty and climate; (4) sport, recreation, and educational facilities; (5) attitudes toward tourists; (6) accessibility of the region; (7) shopping and commercial facilities; and (8) cultural and social characteristics of host country (Peper, 1972; Hills and Mitchell, 1981; Getz, 1983).

TOURISM MARKETING

Marketing may be defined as a system of interrelated activities designed to develop, price, promote, and distribute goods and services to satisfy the needs of a group of prospective customers (Dalrymple and Parsons, 1986). Tourism marketing cannot operate independently of this definition and must be based upon a well-planned program involving the marketing mix for the product of "travel." In order to best attract and serve tourists, tourism marketing professionals must appreciate the importance of the relationship between the marketing concept and travel product planning, including pricing, promotion, distribution, product differentiation, and market segmentation (Melntosh and Goeldner, 1984).

Tourism is a profitable industry for the developed nations. Earnings from tourism have been steadily increasing, accounting for a respectable portion of the GNP in those countries. In the developing nations, on the other hand, tourism marketing is neglected and an important source of income ignored. Tourism revenues in developing countries fall well below the tourism income of developed nations.

The major problem for the developing nations is initial investment opportunities. A tourism industry requires more start-up money than most developing nations have available. In addressing this problem, some countries look for joint venture opportunities with well-known multinational firms. In fact, if a joint venture opportunity is not available or suitable to a developing country's needs, its chances of success in the global tourism market are very small. Jenkins (1982) supported the internationalization of the tourism industry and argued that in the early stages developing nations should seek ventures with multinational chains to develop their tourist products and managerial skills. However, Kaplan (1979) compared the characteristics of a luxury multinational hotel chain with those of a small local hotel complex and concluded that in the long run, establishing a small local hotel complex seems to be more beneficial to a host country in generating revenues, providing employment, and stimulating industrial development.

TOURIST DEMAND FOR TRAVEL TO TURKEY IN THE COMPETITIVE ENVIRONMENT OF THE MEDITERRANEAN BASIN

The rim of the Mediterranean basin is largely developing nations, and as they enter the market for tourism, there is increased competition among them for tourist dollars. In such a competitive market structure, a marketing strategy based on production-era mentality is no longer sufficient. Rather, focus must shift from selling the product/service produced to producing the product which the consumer wishes to purchase. This marketing concept implies that a marketing strategy must be developed through study of the desired target market. To this end, we can make some strategic policy statements, considering our information about the volume of tourists and not on the type of tourist traveling to Turkey. In particular, we find that while the overall growth of tourism has been steady and relatively uninterrupted, tourism remains very seasonal, implying periods of idle capacity and unemployment. Some evidence also suggests that while Turkey may have increased the penetration of its traditional markets, it has not been able to achieve a change in the rate of growth of tourism that would more likely be associated with market development, or diversification.

In examining the basin as a whole, we find that tourist demand by country can be volatile. There is evidence to suggest that while demand for the region's tourist services has grown, the relative market shares of the countries within the basin are not stable. There is also some tentative evidence that the countries' tourist demand can be grouped as complements to and substitutes for each other's tourist demand.

Turkey has a rich culture and history to offer visitors. Its geographic location offers an ideal climate for all types of recreational activities. Its historical epic includes the Hittites, Lydians, Forians, Persians, Greeks, Romans, Byzantines, Seljuks, and Ottomans. It is peninsular, with excellent harbors for both small and large craft. Also, it is a desirable tourist destination during all four seasons. For the purposes of this study, Turkey has some characteristics which make it a particularly useful example: Turkey is an industrialized, developing nation with a well-developed infrastructure; it has a well-documented data history on its tourist industry from which reasonable empirical estimates can be made; like most developing nations, it has experienced some political turmoil and social unrest; it has a stated policy directed to the development of its tourist industry.

Many of Turkey's competitors share some of these characteristics. Italy, Spain, Greece, and Egypt are well-known as popular tourist locations for sightseeing and historical antiquities. Greece has long been a popular boating center. Egypt, Tunisia, Morocco, and Spain all have developed enclave-type tourist centers that offer substantial recreational facilities. Spain, Greece, and southern Italy all have approximately the same level of industrial development as Turkey and, therefore, the same infrastructural base.

Despite the attraction to tourists for its history and its endowments by nature, and despite a policy which stresses the development of a tourist trade, Turkey enjoys only a small share of the basin tourist market. The growth of the industry has generally been positive, however, in spite of fluctuations in demand from time to time. Table 18.1 shows the total number of foreign visitors between 1960 and 1985.

The fluctuations in the number of tourists visiting Turkey are associated with a general disruption in the growth of the Turkish economy, which began in the 1970s. The mid- to late 1970s was a period of political turmoil and social unrest, and as a result, during these years all economic activity suffered. These troubled years ended in a military takeover in 1980.

Beginning with the recovery of the Turkish economy, there was an effort on the part of the Turkish government to increase Turkey's share of the world tourist market. Hilton and Sheraton both have hotel complexes in Istanbul. These two corporations alone plan to open additional complexes in Ankara, Antalya, and Izmir. It is expected by the end of the 1980s that they will be joined by other major resort developers and that the total amount of hotel accommodations will double with this additional development (Etingu, 1984; Economic Dialogue: Turkey, 1984, p. 156). However, even

TABLE 18.1. Total Number of Foreign Visitors to Turkey

Year	Number of Visitors
1960	124,228
1965	361,758
1970	724,784
1975	1,148,611
1980	1,288,060
1982	1,391,717
1983	1,625,099
1984	1,900,000
1985	2,100,000

Source: Statistical Yearbook of Turkey, 1960-1985

the more recent efforts in tourism development seem to employ a marketing strategy that emphasizes accommodating tourists already planning to come to Turkey, rather than actively marketing the advantages of visiting Turkey to a completely untapped market.

EMPIRICAL TEST FOR A STRUCTURAL CHANGE IN THE DEMAND FOR TRAVEL TO TURKEY

The empirical analysis is composed of two different tests. First, a Box-Jenkins time series test determines if there has been a structural change in the process of generating tourists' demand for travel to Turkey. Second, a market share analysis determines to what extent, if any, Turkey's market share of tourists' demand for travel to the Mediterranean basin has changed.

The data for the Box-Jenkins analysis are number of arrivals of tourists by month for the period of 1958 to 1985. These statistics are published in the Statistical Yearbook of Turkey. A Box-Jenkins model was fitted to the entire series and to a subperiod within the series. We conclude that there is no evidence that the disruption of the 1970s or subsequent policy has had any effect on the development of demand for travel to Turkey.

Univariate Box-Jenkins models have been successfully used in other studies to forecast tourists' demand (Geurts and Ibraham, 1975). We propose to use the Q-statistic from Box-Jenkins analysis to construct a F test for a structural change. The estimation of the full period sample is presented in Table 18.2.

The years beginning with 1974 and ending in 1980 were initially hypothesized as those which might be statistically different from the processes' previous experience. Not only were these years known to be those in which political unrest was most intense, but examination of the annual volume of arrivals of tourists indicated a sharp decline in 1974. Additionally, the series appeared to be more volatile during this period than in any other period. A separate model was estimated for this period and is presented in Table 18.3.

A quick examination and comparison of Table 18.2 and Table 18.3 show that the coefficients for both the AR and MA processes are relatively smaller for the 1974-80 subsample. In fact, the coefficient on the AR process is no longer statistically significant. The seasonal coefficient, however, is relatively larger for the subsample. The question thus raised is whether or not these differences are statistically significant.

The test we employed to examine these processes for statistically significant differences in the parameters is an F test constructed from the ration of the Q-statistics. The Q-statistic is distributed as a

TABLE 18.2. Model Estimation and Forecast—Full Sample

	ARIMA (1 1 1) (0 1 1) 12 Parameter Estimates	T-ratio
AR1	.6122	12.69
MA1	.9741	263.06
SMA12	.5962	11.88
Q=5.448		$x^2.05(26)=38.9$
±2SE=.1368		accept H_o
		residual ACF
		is white noise

TABLE 18.3. Model Estimation – Subsample 1974-1980

ARIMA (1 1 1) (0 1 1) 12	Parameter Estimates		T-ratio
AR1	.2586		1.83
MA1	.8896		23.23
SMA12	.7537		6.43
Q=3.726		$x^2.05(18)=28.9$	
		accept H	
		residual ACF	
		is white noise	

chi square, and the F-statistic is a ratio of two chi squares. Thus, the Q-statistic was used from a subsample and from the total sample to construct a one tailed F test. The null hypothesis is that there is no significant difference between the Qs for the total sample and the subsample. Alternatively, one might state this hypothesis as there is no statistical difference between the Qs for the total sample and the subsample. Letting Qa represent the Q-statistic for the whole sample and Qt represent the Q-statistic for the total period, the alternative hypothesis is that Qa is significantly different from Qt. If the calculated value of F, for numerator and denominator degrees of freedom, is less than the critical value of F, one would accept the null hypothesis and conclude that the variations in the ACF of the residuals are equivalent. The implication is that the estimated processes are the same of the two time periods. If the calculated value of F is greater than the critical value of F, one would accept the alternative hypothesis and conclude that the degree of residual variation in the ACF is different and that the processes are different (Isley and Schwer, 1982).

$$F^*_{v1,v2} = \frac{Qa/v1}{Qt/v2} = \frac{.207}{.14005} \qquad Qa = x^2\ .05 \quad Qt = x^2\ .05$$

$$F^*_{18,26} = 1.478$$

F .05 for df 18,26 is 2.05.

Since F* is less than F critical, we reject our hypothesis that the period 1974-1980 is a statistically different period. We conclude that the demand for travel to Turkey is driven by the same generating process throughout the entire 1958 to 1985 time series. Neither the political disruption nor the stated policy to develop the tourist industry has altered tourists' demand.

EMPIRICAL EVIDENCE ON COMPLEMENTARITY AND SUBSTITUTABILITY OF TOURIST DEMAND FOR SERVICES ON THE MEDITERRANEAN BASIN

Our data for the market share analysis cover a span of twenty-one years, 1961-1981, for twelve countries, in addition to Turkey. The countries included are: Spain, France, Italy, Greece, Israel, Yugoslavia, Egypt (data not available until 1970), Algeria (data not available until 1968 and not available for 1974), Morocco, Cyprus, Malta, and Tunisia. The data on annual number of arrivals of tourist were collected from each country's statistical yearbook.

The covariance analysis must be viewed as a very crude approach to the question of substitutability or complementarity. (The correlation coefficient is actually used and the sign is retained from the covariance in order to measure direct and indirect relationships.) It is felt that if data on the number of tourist arrivals where other travel stops are included in the tourist's package were available a most direct test would be conducted. The evidence presented below is limited by what data are readily available. However, given that one can possibly identify those tourist services which are viewed as complements and substitutes, Turkey could devise a strategy for both diversification and market development. Development of complementarity with successful national tourist industries would provide a focus for new market development.

Simultaneously, to the extent that additional data would identify those countries whose tourist industry is viewed as a close substitute, a reasonable marketing strategy would suggest that Turkey should seek to significantly differentiate, or diversify, its product from those which it is most like, and address the uniqueness of travel to Turkey.

RESULTS OF MARKET SHARE

The focus of the market share analysis was limited to the Mediterranean basin because it was felt that the tourist industries of these countries would be viewed by travelers as close substitutes and/or complements for travel to Turkey. If these other national tourist services are substitutes, to the extent that they have significantly changed the desirability of their tourist services, their share of travel demand would increase at the expense of the demand for travel to Turkey, suggesting one should look for an inverse relationship. If the services of these other national tourist industries are viewed as a complement, then to the extent that other tourist industries have improved their share or lost their share, travel to Turkey should show movements in the same direction, and one should look for a direct relationship. The following is a simple trend analysis of market share and a covariance analysis to indicate direct or inverse relations between market share. Table 18.4 is the summary statistics on the market by country. Market share data are in Appendix A. Appendix B is the table of correlation coefficients. Any correlation coefficient that is slightly larger than 1.351 will pass a t test for significance at x = .05.

Time series plots of the market shares of each of these thirteen countries (see Table 18.4) indicated that Greece, Israel, Egypt, Malta, and Tunisia all show a very strong and almost consistently positive trend in marketing their market share. Spain and Morocco have both shown a negative trend in market share since 1974, and Yugoslavia has shown a negative trend in market share since 1970. France has shown a highly volatile market share over time, with a boom ending in 1968, the bottom of that cycle ending in 1974, and a recovery for the most part since 1974. Algeria showed two periods of growth in market share, one between 1968 and 1971, and

TABLE 18.4. Percent Share of Total Travel to Thirteen Mediterranean Basin Countries

Country	Mean	Standard Deviation
Spain	$\bar{x}$=.34	σ =.10
France	$\bar{x}$=.25	σ =.05
Italy	$\bar{x}$=.26	σ =.08
Greece	$\bar{x}$=.03	σ =.01
Turkey	$\bar{x}$=.01	σ =.006
Israel	$\bar{x}$=.008	σ =.002
Yugoslavia	$\bar{x}$=.06	σ =.03
Egypt	$\bar{x}$=.01	σ =.002
Algeria	$\bar{x}$=.003	σ =.001
Morocco	$\bar{x}$=.01	σ =.004
Cyprus	$\bar{x}$=.01	σ =.02
Malta	$\bar{x}$=.003	σ =.002
Tunisia	$\bar{x}$=.001	σ =.005

another between 1973 and 1976. The second period of growth for the most part was a recovery of the market share in the 1968 through 1971 period. In 1977 Algeria sank below its 1968 market share and has shown virtually no growth since. Cyprus showed almost no growth at all until 1978 but since then has generally held about a 6% market share. Finally, Italy showed a decline in market share until 1975 and has shown only modest gains in market share since that time.

Turkey generally showed gains in market share until 1977 and since that time has shown a decline. Although the downturns are not exactly coincided, the pattern of market share for Turkey most resembles that of Spain, Morocco, and Yugoslavia.

An analysis of the correlation (covariance) shows a strong positive or direct relation between the Turkish market and market shares for Spain, Greece, Egypt, Morocco, and Tunisia. It is hypothesized, therefore, that travel to these countries complements travel to

Turkey. While the growth pattern for Yugoslavia suggests that it should probably also appear in this group, the correlation (covariance) analysis does not confirm this. It should be noted that the correlation coefficient between the market share for Turkey and for Yugoslavia is very small (.02) so that the results for Yugoslavia on the correlation analysis are not statistically significant. Among countries with a market share positively related to the Turkish share, the least correlation is 11%, the largest being Spain with 20%. Note, however, that with a standard t test, no correlation is statistically significant.

There is an inverse relation between the Turkish market share and the market shares for France, Italy, Israel, Algeria, Cyprus, and Malta (and Yugoslavia, as noted above). It is hypothesized that these countries lure tourists who may consider travel to Turkey a close substitute. Italy's correlation is the largest, at 22%, and the smallest is Malta's at 1%. Again, no correlation is statistically significant.

Our conclusions are drawn on rather weak grounds, both in terms of the types of test upon which we rely, and second, of the relations discovered that are not statistically significant. It may be just as possible on the basis of the data to argue that demand for Turkish tourist services is not part of the same market as the Mediterranean basin.

CONCLUSIONS

This study of tourists' demand for travel to Turkey suggests that the Turkish government's current industrial policy has been ineffective in maximizing the rate of growth in the tourist industry. Between 1958 and 1985, tourists' demand for travel to Turkey shows only a slight positive trend with short-term fluctuations around the growth path. Analysis of the data, in fact, leads to one very obvious policy recommendation. The very dominant seasonal pattern implies that physical facilities have excess capacity during large portions of the year. This pattern also implies that labor in the tourism sector will be only seasonally employed. Most tourist-based economies do suffer seasonality. However, because of the associated costs, there has been in recent years a strong emphasis on encouraging "year-round resort" types of developments. Such diversification could raise the volume of tourists with minimal additional investment.

There are clearly a number of limitations to the analysis of market shares. First, very few of the countries showed an uninterrupted pattern of growth. The time series plots clearly indicate structural change in the pattern of market shares. Because of the very limited number of observations, it was felt that curve-fitting exercises would be of limited value, given so few degrees of freedom. The reliability of plots is, however, clearly limited. What is indicated is that the pattern of change in market share for Turkey is not unique, and therefore external forces may well be at work: forces not entirely internal, relating to competition among the region's various national tourist industries, but external to the region, part of a general change in tourist demand.

The correlation analysis does, however, indicate some pattern of regional interdependence, which may be loosely described in terms of complementarity and substitutability of tourist demand for travel in the region. Combining the correlation results with those of the time series plot, the marketing strategy suggested is more complex than that hypothesized. A case may well be made that an effective marketing strategy should be worked out among those nations having complementary types of demand. Such a strategy might include package tours highlighting Ottoman influence in architecture or culture, mutually increasing the demand for all participant countries. These countries are currently, collectively, losing their share to their competitors. Simultaneously, since their competitors, offering closely related travel services, have collectively been clearly more effective in improving market share, a strategy might be pursued, in which Turkey is made to look more like those countries which serve as close substitutes. On the other hand, Turkey can attract the allocentric and the near-allocentric segment, as well as mid-centric segments, and this in effect suggests that Turkey should differentiate its product from those in its current market group and begin to adopt a strategy more like the Greece, Egypt group. It should be noted that these two strategies are not mutually exclusive. A multiple-target strategy would combine these two strategies.

The recommendations for competitive marketing strategies for Turkey, therefore, are: (1) It must be recognized that there are a large number of close substitutes for travel to Turkey, since Turkey shares the Mediterranean and Aegean basin with Greece, Yugoslavia, Italy, France, Spain, and North Africa. (2) Tourists tend to seek travel to sites which offer a "bundle" of recreational opportunities. This implies that tourists want a variety of sports facilities, and dining, sightseeing, and shopping opportunities. Turkey must represent a genuine qualitative alternative. (3) The government should not only publicize the need to be friendly and helpful to foreign visitors, but should also educate local residents on how to assist visitors in a manner which minimizes cultural misunderstandings. This should include language assistance on signs and menus and some interpretational assistance at public locations. (4) A comprehensive marketing strategy and associated advertising plan should be developed and implemented. This should begin with a reevaluation of Turkey's comparative advantages and should include a forecast of both quantity and types of tourists attracted by alternative forms of recreation and facilities.

REFERENCES

Bond, M.E. & Ladman, J.R. (1972). "Tourism: A Strategy for Development." *Nebraska Journal of Economics and Business*, Vol. 11 Winter, pp. 37-52.

"Come and See for Yourself" (1984). *Economic Dialogue: Turkey*, Special Issue, January, p. 156.

Dalrymple & Parsons (1982). *Marketing Management*, Wiley.

Development Plan of Turkey: 1985-1989 (1985). Ankara: State Planning Organization.

Diamond, J. (1977). "Tourism's Role in Economic Development: The Case Reexamined." *Economic Development and Cultural Change*, Vol. 25, No. 3, pp. 539-553.

Diamond, J. (1974). "International Tourism and the Developing Countries: A Case in Failure." *Economica Internationale*, Vol. 27, August-November, pp. 601-615.

Etingu, Ali (1984). "A Comment on Turkish Tourism." *Economic Dialogue: Turkey*, Special Issue January, pp. 138-139.

Getz, Donald (1983). "Capacity to Absorb Tourism: Concepts and Implications for Strategic Planning." *Annuals of Tourism Research*, Vol. 10, No. 2, pp. 239-263.

Geurts, M.D. & Ibraham, I.B. (1975). "Comparing the Box-Jenkins Approach with the Exponentially Smoothed Forecasting Model: An Application to Hawaii Tourists." *Journal of Marketing Research*, Vol. XII, May, pp. 182-188.

Ghali, M. (1977). *Tourism and Regional Growth*, Martinus Nijhoff Social Sciences Division.

Hills, P.J. & Mitchell, C.G.B. (1981). "New Approaches to Understanding Travel Behavior," in *New Horizon in Travel-Behavior Research*, P. R. Stephen, A. H. Meyburg, and W. Grog (eds.), Lexington, Massachusetts: Lexington Books.

Isley, P.W. & Schwer, R. Keith (1982). "Forecasting the Demand for Scrap Steel," Working Paper, Norwich University Workshop Series.

Jenkins, C.L. (1982). "The Effects of Scale in Tourism Projects in Developing Countries." *Annals of Tourism Research*, Vol. 9, No. 2, pp. 229-249.

Jenkins, C.L. (1982). "The Use of Investment Incentives for Tourism Projects" *International Journal of Tourism Management*, Vol. 3 June, pp. 91-97.

Kadt, Emanuel D. (1979). *Tourism: Passport to Development*. Oxford University Press.

Kaplan, A. (1979). "When Less is More: A Look at the Long Term Building for Tourism." *The Cornell Hotel and Restaurant Administrative Quarterly*, Vol. 20, August, pp. 4-5.

Kotler, P. (1988). *Marketing Management* (6th ed.). Prentice-Hall.

Kyle, P.W. (1978). "Lydia Pinkham Revisited: A Box-Jenkins Approach," *Journal of Advertising Research.* Vol. 18, No. 2, April, pp. 31-38.

Melntosh, R.W. & Goeldner, C.R. (1984). *Tourism: Principles, Practices, Philosophies* (4th ed.). Grid Publishing.

O'Connor, T.S. (1981). "Marketing to the International Visitor." *The Cornell H.R.A. Quarterly*, Vol. 15, February, pp. 53-59.

Pearce, P.L. (1982). *The Social Psychology of Tourist Behavior*. Pergamon Press.

Peper, H.W.T. (1972). "Tourism in Developing Countries: Some Economic and Fiscal Considerations." *Bulletin for International Fiscal Documentation*, Vol. 24, April, pp. 147-159.

Reime, M. & Hawkins, C. "Tourism Development: A Model for Growth." *The Cornell H.R.A. Quarterly*, Vol. 13, May, pp. 67-74.

Rovelstad, J.M. & Blazer, Suzanne R. (1983). "Research and Strategic Marketing in Tourism: A Status Report." *Journal of Travel Research*, Vol. 22, No. 2, Fall, pp. 2-7.

Statistical Yearbook in Turkey, (1965-1982). Ankara: State Institute of Statistics.

Theobald, T.C. (1984). *Turkey: A Model for Adjustment Programs*, Citicorp, Public Affairs Division.

Tourism Policy and International Tourism in OECD Member Countries, (1983). OECD Publication.

Turizm Istatistikleri Bulteni, (1984). Kultur ve Turizm Bakanligi.

Turner, L. (1976). "The International Division of Leisure: Tourism and the Third World." *World Development*, Vol. 4, March, pp. 253-260.

Wells, R.J.G. (1982). "Tourism Planning in a Presently Developing Country: The Case of Malaysia." *International Journal of Tourism Management*, 3, June, pp. 98-107.

Willis, F.F. (1977). "Tourism as an Instrument of Regional Economic Growth." *Growth and Change*, Vol. 8, April, pp. 43-47.

Wood, R.E. (1980). "International Tourism and Cultural Change," *Economic Development and Cultural Change*, Vol. 28, No. 3, pp. 561-581.

APPENDIX 18A. Market Share by Countries in the Mediterranean Basin 1961-1981

ROW	SPAIN	FRANCE	ITALY	GREECE	TURKEY	ISRAEL
1	0.271618	0.249402	0.404326	0.0152800	0.0041810	0.0050652
2	0.265364	0.243944	0.403679	0.0185145	0.0054298	0.0061364
3	0.281141	0.229083	0.394904	0.0207612	0.0075990	0.0063376
4	0.313633	0.286905	0.304105	0.0193059	0.0557037	0.0056893
5	0.362929	0.263777	0.274328	0.0173346	0.0032708	0.0064748
6	0.345628	0.269199	0.269199	0.0205416	0.0086144	0.0072005
7	0.342672	0.255349	0.274825	0.0215878	0.0082664	0.0062712
8	0.174652	0.322901	0.339046	0.0228452	0.0093022	0.0078357
9	0.186735	0.291334	0.339890	0.0237249	0.0103370	0.0116615
10	0.401151	0.238645	0.191145	0.0224721	0.0085735	0.0080666
11	0.459244	0.261007	0.197561	0.0268151	0.0085046	0.0079826
12	0.461956	0.253782	0.181024	0.0342053	0.0159866	0.0106623
13	0.536895	0.163122	0.181313	0.0402407	0.0170946	0.0012279
14	0.478925	0.144488	0.182349	0.0394550	0.0185980	0.0083705
15	0.447202	0.187131	0.183373	0.0288280	0.0163638	0.0084008
16	0.424704	0.184189	0.186586	0.0400411	0.0217265	0.0078813
17	0.411066	0.184483	0.190783	0.0526604	0.0229542	0.0100390
18	0.236282	0.295521	0.208153	0.0445672	0.0142669	0.0100588
19	0.265006	0.289202	0.206759	0.0488215	0.0131641	0.0103309
20	0.248267	0.297538	0.226730	0.0541326	0.0109341	0.0104376
21	0.228838	0.306134	0.225725	0.0487780	0.0091738	0.0108418

ROW	YUGOSLAV	EGYPT	ALGERIA	MOROCCO	CYPRUS	MALTA	TUNSIA
1	0.038793	0.0000000	0.0000000	0.0070290	0.0010486	0.0008797	0.0023771
2	0.045403	0.0000000	0.0000000	0.0067673	0.0017749	0.0009590	0.0019389
3	0.047615	0.0000000	0.0000000	0.0077179	0.0019247	0.0008933	0.0020244
4	0.050343	0.0000000	0.0000000	0.0082399	0.0021403	0.0009267	0.0030068
5	0.057310	0.0000000	0.0000000	0.0096143	0.0004147	0.0009882	0.0035591
6	0.064455	0.0000000	0.0000000	0.0091770	0.0008052	0.0011593	0.0040210
7	0.074369	0.0000000	0.0000000	0.0091753	0.0011707	0.0015775	0.0047348
8	0.098969	0.0000000	0.0030003	0.0107661	0.0018405	0.0026236	0.0062185
9	0.104864	0.0000000	0.0034825	0.0129779	0.0023873	0.0036956	0.0089100
10	0.093610	0.0068103	0.0042029	0.0119638	0.0023273	0.0036704	0.0073625
11	0.000730	0.0068147	0.0044943	0.0133552	0.0024119	0.0032559	0.0078245
12	0.001093	0.0073908	0.0030178	0.0142135	0.0030834	0.0030851	0.0105000
13	0.001250	0.0089338	0.0032488	0.0175388	0.0037707	0.0024758	0.0128895
14	0.085229	0.0074115	0.0000000	0.0185842	0.0036586	0.0029241	0.0100058
15	0.080441	0.0100146	0.0036698	0.0177596	0.0022255	0.0040383	0.0105526
16	0.082268	0.0111819	0.0041874	0.0175532	0.0006627	0.0047232	0.0142964
17	0.076313	0.0134767	0.0025337	0.0151750	0.0024652	0.0046566	0.0133945
18	0.068814	0.0112965	0.0027229	0.0160672	0.0767465	0.0040730	0.0114315
19	0.068783	0.0113328	0.0028009	0.0159112	0.0504374	0.0051493	0.0123023
20	0.061715	0.0110065	0.0027516	0.0148547	0.0412124	0.0063929	0.0140271

APPENDIX 18B. Correlation Coefficient of Market Share of Tourist Demand with Covariance Signs for Thirteen Mediterranean Basin Countries

	SPAIN	FRANCE	ITALY	GREECE	TURKEY	ISRAEL	YUGOSL	EGYPT	ALGERIA	MOROCCO	CYPRUS	MALTA
FRANCE	-0.797											
ITALY	-0.683	0.357										
GREECE	0.096	-0.120	-0.657									
TURKEY	0.197	-0.134	-0.215	0.140								
ISRAEL	-0.348	0.376	-0.278	0.477	-0.077							
YUGOSL	-0.431	0.110	0.060	0.042	-0.018	0.416						
EGYPT	0.310	-0.264	-0.804	0.906	0.131	0.426	-0.010					
ALGERIA	0.172	0.011	-0.588	0.458	-0.028	0.402	0.041	0.635				
MOROCCO	0.442	-0.420	-0.842	0.773	0.157	0.349	0.096	0.844	0.643			
CYPRUS	-0.429	0.450	-0.259	0.641	-0.059	0.474	0.090	0.552	0.221	0.348		
MALTA	-0.091	0.116	-0.595	0.840	-0.011	0.683	0.252	0.845	0.693	0.704	0.618	
TUNISIA	-.915	-0.158	-0.757	0.911	0.106	0.488	0.091	0.917	0.696	0.883	0.515	0.910

Index